SOCIAL DECISION METHODOLOGY
FOR TECHNOLOGICAL PROJECTS

THEORY AND DECISION LIBRARY

General Editors: W. Leinfellner and G. Eberlein

Series A: Philosophy and Methodology of the Social Sciences
Editors: W. Leinfellner (Technical University of Vienna)
G. Eberlein (Technical University of Munich)

Series B: Mathematical and Statistical Methods
Editor: H. Skala (University of Paderborn)

Series C: Game Theory, Mathematical Programming and Operations Research
Editor: S. H. Tijs (University of Nijmegen)

Series D: System Theory, Knowledge Engineering and Problem Solving
Editor: W. Janko (University of Economics, Vienna)

SERIES A: PHILOSOPHY AND METHODOLOGY OF THE SOCIAL SCIENCES

Volume 9

Editors: W. Leinfellner (Technical University of Vienna)
G. Eberlein (Technical University of Munich)

Editorial Board

Scope

This series deals with the foundations, the general methodology and the criteria, goals and purpose of the social sciences. The emphasis in the new Series A will be on well-argued, thoroughly analytical rather than advanced mathematical treatments. In this context, particular attention will be paid to game and decision theory and general philosophical topics from mathematics, psychology and economics, such as game theory, voting and welfare theory, with applications to political science, sociology, law and ethics.

For a list of titles published in this series, see final page.

t54.00

SOCIAL DECISION METHODOLOGY FOR TECHNOLOGICAL PROJECTS

edited by

CHARLES VLEK

University of Groningen, The Netherlands

and

GEORGE CVETKOVICH

Western Washington University, Bellingham, U.S.A.

KLUWER ACADEMIC PUBLISHERS

DORDRECHT / BOSTON / LONDON

ISBN 0-7923-0371-7

Published by Kluwer Academic Publishers,
P.O. Box 17, 3300 AA Dordrecht, The Netherlands.

Kluwer Academic Publishers incorporates
the publishing programmes of
D. Reidel, Martinus Nijhoff, Dr W. Junk and MTP Press.

Sold and distributed in the U.S.A. and Canada
by Kluwer Academic Publishers,
101 Philip Drive, Norwell, MA 02061, U.S.A.

In all other countries, sold and distributed
by Kluwer Academic Publishers Group,
P.O. Box 322, 3300 AH Dordrecht, The Netherlands.

printed on acid free paper

Printed in The Netherlands

PREFACE

This book grew out of the conviction that the preparation and management of large-scale technological projects can be substantially improved. We have witnessed the often unhappy course of societal and political decision making concerning projects such as hazardous chemical installations, novel types of electric power plant or storage sites for solid wastes. This has led us to believe that probabilistic risk analysis, technical reliability analysis and environmental impact analysis are necessary but insufficient for making acceptable, and justifiable, social decisions about such projects. There is more to socio-technical decision making than applying acceptance rules based on negligeably low accident probabilities or on maximum credible accidents. Consideration must also be given to psychological, social and political issues and methods of decision making.

Our conviction initially gave rise to an international experts' workshop titled 'Social decision methodology for technological projects' (SDMTP) and held in May 1986 at the University of Groningen, the Netherlands, at a time when Cvetkovich spent a sabbatical there. The workshop - aimed at surveying the issues and listing the methods to address them - was the first part of an effort whose second part was directed at the production of this volume. Plans called for the book to deal systematically with the main problems of socio-technical decision making; it was to list a number of useful approaches and methods; and it was to present a number of integrative conclusions and recommendations for both policy makers and methodologists.

The University of Groningen workshop on SDMTP was attended by about 20 technological and behavioral decision theorists and a smaller number of strategic planners and policy makers. A volume of specifically prepared working papers had been distributed, and the three workshop days were scheduled to accommodate leading introductions, commentaries by assigned discussants, and 'summaries of the day' aimed at an orderly recapitulation and integration of the main issues, comments and suggestions. The summary notes prepared by B. Brehmer and G. Keren (day 1), P. Koele and L. Phillips (day 2) and K. Begg and P. Neijens (day 3), have been particularly useful in the preparation of the concluding chapter.

In the wake of this meeting, the initial working papers were carefully reviewed and their authors invited to prepare and submit a publishable chapter that would fit into the planned book. This succeeded in most cases. Also, several missing links in the sequence of envisaged papers were forged after the fact, by inviting additional, external authors to contribute a chapter.

After all revised chapters had been submitted, we have scrutinized them once more, in order to check for mutual consistency and to insert cross-references between chapters. Finally, we have completed an 'introduction and overview' (first chapter) and a 'review of key issues and a recommended procedure' (final chapter), in which we have attempted to provide an embedding structure for the more specific chapters in between.

The University of Groningen has kindly made available the space and facilities needed to conduct the workshop and prepare this book. The Netherlands Science Organization (N.W.O.) and the Ministry of Science and Education (Directorate-general of Science Policy) have generously supported our efforts financially. Earlier versions of the various chapters have been critically reviewed and commented upon by K. Begg (London), Th. Bezembinder (Nijmegen), H. Boer (Enschede), H. Brouwer (The Hague), M. Charles (Fort Wayne, Ind.), T. Earle (Seattle), P. Koele (Amsterdam), G. Keren (Soesterberg), P. Lourens (Groningen), C. Midden (Leiden), P. Neijens (Amsterdam), F. Siero (Groningen), Tj. van der Schaaf (Eindhoven) and W.A. Wagenaar (Leiden).

Ms. Magda Bootsma and ms. Ineke Teunis-Vos of the Department of Psychology in Groningen have carefully assisted in preparing parts of the manuscript. Our junior colleague, drs. Wilma Otten, has provided us with essential substantive and text-processing help in the final stage of preparing the camera-ready manuscript for the publisher.

To these various organizations and the many persons upon whom we have been dependent during the preparation of this volume, we express our sincere thanks for their encouragement and support, their patience and their loyal assistance.

Groningen/Bellingham,
November 1988   Charles Vlek   George Cvetkovich

CONTENTS

INTRODUCTION AND OVERVIEW:
SOCIAL DECISION MAKING ON TECHNOLOGICAL PROJECTS

CHARLES VLEK AND GEORGE CVETKOVICH

Department of Psychology
University of Groningen

Department of Psychology
Western Washington University, Bellingham

Suppose your local government is approached by your Ministry of Energy with the proposal to site a new coal-fired electric power plant within its jurisdiction. Or assume that the regional airport in your vicinity is being considered for significant expansion and more intense utilization, especially for handling freight planes by night. Or imagine that your regional authorities get the idea of routing a regular heavy transport of radioactive waste along a main thoroughfare in your town.

Due to the enlarged scale of such technological projects, and stimulated by published failure incidents and accidents, public evaluation and decision making have become a burdening and sometimes painful experience for the responsible authorities. In the 1970's Japan had its revolt of regional peasants supported by activists from Tokyo, against the government's plans to expand Narita International Airport. The Federal Republic of Germany witnessed its 'green' uprisings about the plan to construct a new runway for the airport of Frankfurt. North-West Canada had its unrest and deliberations about the U.S. government's idea to build the Mackenzie Valley pipeline, channeled in the Berger Inquiry (Gamble, 1978). Safety regulations and routing of liquefied energy gas transports were heavily discussed after the catastrophic BLEVE (Boiling Liquid Expanding Vapour Explosion) near the populated Los Alfaques campground in Spain (1979).

But the limitations of technological analysis and decision making have been most heavily debated in the area of nuclear electric power generation and the storage of radioactive waste. Austria's expert discussion sessions and subsequent referendum led to the closure of its only, nearly finished nuclear power plant at Zwentendorf (Nelkin and

1

Ch. Vlek and G. Cvetkovich (eds.), Social Decision Methodology for Technological Projects, 1–13.
© *1989 by Kluwer Academic Publishers.*

Pollak, 1978). Sweden's public debate and referendum (Abrams, 1979; Nelkin and Pollak, 1978) yielded the decision that their twelve nuclear power plants will be utilized until they are worn out, and that new ones will not be built. Great Britain had its Windscale (now Sellafield) Inquiry (Parker, 1978) - a piece of political ritual according to Wynne (1982), and later, its Sizewell B Inquiry in which serious, but not fully fledged examinations were made into the concept of acceptable (public) risk (O'Riordan, Kemp and Purdue, 1985). The F.R. of Germany had its 'Enquête-Commission' concerning future electric energy scenarios. Later the F.R.G. had to face its public protests and legal blockage of the development of an underground storage facility for radioactive waste near the town of Gorleben. Recently, in the U.S.A. a long project has been concluded concerning the permanent underground storage of radioactive waste (see, e.g., Gregory and Lichtenstein, 1987). Canada put its Porter Commission to work (Royal Commission, 1980), which - through research, expert consultation and public hearings - had to design a feasible 'electric energy future' for the province of Ontario.

The small country of The Netherlands has played it big by organizing, in 1981-1983, a government-sponsored 'Wide Societal Discussion on Energy Policy' (see Jansen, 1985; Vlek, 1986), which yielded the - reserved - conclusion that, in view of the unsolved storage problem of radioactive waste and strong public opposition, "it would not be obvious, now, to go ahead and build additonal nuclear power plants". One year later, the citizenry witnessed a stubborn government's 'principal decision' for the construction of at least two and perhaps four new nuclear power stations, which had to be withdrawn - "for the time being" - after the truly continental accident at Chernobyl. The Dutch radioactive waste storage policy is a fairy tale in itself, told by Brouwer in the present volume.

"Due to the enlarged scale of such technological projects...", we said above. What are the characteristics of Large-Scale technology (LST) that often make it so controversial, so difficult to grasp and so hard to decide about? Here is a short list of proposed answers:

- LST is aimed at massive benefits (like cheap electricity, disposal of hazardous waste, more efficient transport), but it may also have serious negative consequences. This makes it at best a dilemma for those who need and/or appreciate the benefits; for others, outright rejection is a more likely response.

- LST, therefore, is in the interests of special (maybe large) groups of people, but it may ultimately affect many more (and larger) groups of people. Thus responsible decision makers have to take notice of a variety of interests, goals and evaluations, and problems of fairness may easily arise.

- LST can be reasonably well understood, evaluated and controlled by relatively few people, but its actual consequences will, or may, affect a multitude of people. This creates a knowledge gap. For many the technology may be incomprehensible and uncontrollable, thus requiring an unusual degree of public trust in technology operators and administrators.

- LST has both short-term and long-term consequences, part of which are intended and direct, and part of which are unintended and indirect. A peculiar feature - making LST-evaluators 'motivationally imbalanced' - is that intended benefits are immediate and direct, whereas inevitable and unintended costs and possible harm occur on the long term and have indirect effects ("who would worry about them now?").

- LST is novel; relevant projects tend to be unique. Relatively litte technical experience has accumulated and, in so far as they exist, material failure statistics may be unreliable. Actual consequences have not been experienced either. Hence valid data for judging technical feasibility and for evaluating social acceptability are lacking. Judgment and decision making, therefore, are loaded with ambiguities, uncertainties and complete unknowns.

- LST is complex, i.e., it rests upon a multitude of different subsystems, components and (sub-) processes, each of which demands supervision and control by trained (and motivated) specialists. This complex whole is dominated by the need for strict coordination, and therefore requires some kind of hierarchical organization. The combination of interactive complexity and tight coupling of subsystems and components makes LST vulnerable to small but crucial operating errors that could (and will, says Perrow, 1984) lead to catastrophic accidents.

- The 'catastrophe potential' of LST has apocalyptic proportions. For many people, just mentioning the names of Seveso (dioxine), Torrey Canyon and Amoco Cadiz (crude oil), Harrisburg (nuclear power),

Love Canal (N.Y.; chemical waste), Mexico City (LPG), Bhopal (metylisocyanate), Chernobyl (nuclear power) and Sandoz-Basel (the Rhine pollution disaster) is sufficient to elicit strong memories of frightful events. This raises critical questions about the *a priori* assessment of risks and about criteria and rules for evaluating which risks are 'acceptable' (and for whom). A classical statistical view of risk and risk taking appears too restrictive to deal with these questions in a socially and politically feasible manner.

Through these seven global characteristics LST appears to be bigger than what can be handled by traditional, technical and political decision making procedures. For one thing, the depth of technical feasibility studies has been increased by the development of system reliability and risk analysis methodologies. Also, public concern about LST-projects has been partially met by government requirements that safety reports and environmental impacts statements are to be submitted when initiatiors of a project apply for an operating licence.

However, despite the increased recognition of the need for careful and multifaceted analysis, systematic public involvement in decision making about LST is rare, and it almost seems to be a commodity which policy makers would rather not consider. The assumption is, of course, that - at least in a democratic society - lawful regulations and administrative procedures exist such that - somehow - the *vox populi* is supposed to be reflected in the manner in which an LST-project is officially dealt with.

That assumption, however, does not seem to be valid enough. It would appear that each LST-project brings its own specific relevant interest groups, and its own (potentially) affected publics with it. More often than not these groups do not coincide with the constituencies of elected politicians. Or they may have more accurate senses to gauge the meaning of the possible consequences of the project, than their official representatives who may be taking a more distant perspective. And, of course, *ad hoc* interest groups pertaining to a particular LST-project may not be politically committed at a more general, e.g., organizational or party level, and conversely.

From a methodical point of view, let us briefly consider the problems of social decision making about technological projects. Obviously, and first of all, there is every single involved person's task of forming, for him- or herself, a view of the decision problem and of developing relative preferences about alternative courses of action. This we could label the *individual judgment problem*. Our judge might wish to do this largely on the basis of (somehow) educated intuition. Or (s)he may prefer to evaluate the technological alternatives with the support of plenty of technical, economic, and environmental impacts analyses. Whatever the case may be, the individual judgment process may be understood (and, perhaps, organized) in terms of three subsequent interlocking phases: information acquisition, structuring of the decision and evaluating alternative courses of action. Each of these is complex enough to be broken down in subphases, whose importance depends on the nature of the LST-project under consideration and the educational state of the individual. Various specific methods and techniques may be used to overcome the natural cognitive limitations of individual human decision makers.

Secondly, there is the *social interaction and aggregation problem*: when a number of well-informed, clear-minded individuals is to make a collective decision about an LST-project, this requires a process of: (a) information exchange, (b) joint problem structuring, and (c) collective evaluation, in which different views, judgments and preferences must somehow be accommodated and/or aggregated. Again, these three global phases of the process may be broken down into subphases, each of which may be more or less crucial depending upon the nature of the LST-project and the particular composition of the set of individuals.

In practical cases of social decision making about technological projects, the individual judgment problem and the social interaction and aggregation problem are merged into a diffuse whole: well informed persons debate with persons who "know nothing about it". The well informed and know-nothings may cherish opposing views, or they may, after a while, surprisingly discover that they aspired the same things all along. Factual judgments may be - unconsciously - distorted by goals and interests to which one is committed. Risks may be judged high or low depending on whether one is in a powerless or in a controlling position vis-à-vis the activity under consideration.

The present collection of chapters deals with the question of whether social decision making about technological projects can be orderly structured, and why it should. The contributing authors take the reader along various theoretical and methodological signposts that should be borne in mind by anyone who aspires to design a social decision procedure which is to relieve social and political tension, prevent delay and/or stagnation of projects important to society, and increase the (potentially) affected citizens' sense of involvement. The need for, and the practical use of a structured decision methodology are vividly illustrated in two case studies, one concerning an experimental wind turbine park, the other about a long-term interim storage facility for radioactive waste.

The first chapter to follow is a theoretical essay by *Bezembinder* (Nijmegen) about the impossibility of (purely) rational collective choice. That is, given that each party in a social decision situation has reached a distinct preference order of the available courses of action, there is no *a priori* acceptable rule or system by which the various individual preference orders can be aggregated to form a single collective preference order. So what do you do in practice? You follow a 'reasonable' decision approach. In it you seek to delineate a socially efficient choice set, you attempt to specify 'fairness' ("a human concept") by compromising on egalitarianism, rationality and decisiveness, and you try to elicit social or ethical value judgments rather than purely individual ('egoistic') and/or technical ones.

'Reasonable' collective decision making is necessary because without it society would frequently find itself in an unsolvable social dilemma. *Wilke* (Groningen) deals with such dilemma situations, where separate persons make individually attractive choices whose interdependence may lead to an unwanted social consequence, i.e., a shortage, or a severe scarcity, of an economic or environmental good. Social dilemmas, says Wilke, may be solved by introducing rewards and/or punishments (which would change the structure of the dilemma), by moral appeals, by vividly portraying the unwanted social effects of individual 'greed', and/or by strong leadership. Such measures should be aimed at restoring feelings of fairness in the distribution of the relevant good, since beliefs of unfairness would cause the return of the social dilemma situation.

Brehmer (Uppsala) presents an overview of social judgment theory and analysis. Brehmer's claim is that a fair amount of social controversy over technological projects may be due to cognitive rather than value conflicts. When participants in the debate would analyze their overall judgments in terms of separate underlying components, the sources of conflict and misunderstanding may become visible, and conflict may be considerably reduced. Brehmer also discusses a computer program,

POLICY, which would aid in the identification of individual judgment policies. He recommends that the separation of facts and values in technology debates be achieved by letting different groups of judges deal with them.

A new taxonomy of intra- and interorganizational conflict is proposed by *Vári* (Budapest), in terms of structural levels and functional stages of social decision making. In combination these identify five different types of possible conflict revolving around: (a) problem boundaries, (b) structure of scenarios, (c) scenario data, (d) evaluation structure, and (e) evaluative data, respectively. Using Kilman and Thomas's (1974; see Thomas, 1976) idea that conflicts may be solved along a distributive or along an integrative dimension, Vari then discusses three basic approaches for handling social conflicts: (1) the *reconciling* approach which is basically aimed at conflict avoidance through requisite decision modelling (see Phillips, this volume), (2) the *accommodating* approach which involves bargaining and compromising, and (3) the *confronting* approach in which divergent views are fully exposed and integrated into a higher-level, more-encompassing problem definition and evaluation of proposed alternatives for choice.

Phillips (London) gives an exposé of "decision conferencing", an intensive two-day problem solving session attended by involved decision makers, and structured around the concept of requisite decision model. Applied to evaluating LST-projects, decision conferencing would facilitate problem structuring and communication among interest groups, and it would aim for an effective balance among content, structure and process of decision making. Phillips warns that for a decision conference to be successful, authority and accountability should be well distributed; there should be no concentrations of power.

Value-oriented social decision analysis (VOSDA), explained by *Chen* and *Mathes* (Ann Arbor), differs from decision conferencing in that it emphasizes individual problem clarification and evaluation before the relevant parties get together to discuss their differences and commonalities. Chen and Mathes explicitly recognize the impossibility of (non-dictatorial) rational collective choice (cf. Bezembinder, this volume). Their claim is that VOSDA is emotionally less demanding than face-to-face decision conferencing, and that it establishes trust between parties involved, furthers and explicates the position of interest groups, and provides a 'script' for further negotiation. Thus

VOSDA might well precede a more binding decision conference, and this combination might be very suitable for handling LST decision problems in the public sector.

With a group of colleagues in the Citizen Participation Research Unit at Wuppertal, *Dienel* has field-tested an approach in which representative groups of citizens are asked to commit themselves for a series of meetings in which a proposed public project (a new town-hall, electric power station, freeway crossing or public park) is to be analyzed and evaluated. Each group receives prior information and expert advice, may pay site visits and may achieve a final judgment with the help of maquettes simulating alternative options. The primary outcome of this process is a 'citizen report', one for each group, which is intended to aid responsible decision makers in making a definite choice. This 'planning cell' process might be further improved by some kind of computer support and by application of formal multi-attribute utility analysis (MAUA). Renn, Stegelmann, Albrecht and Kotte (1984) have utilized MAUA in a 'planning cell' study directed at the evaluation of electric power generating scenarios for the Federal Republic of Germany.

The question of possible computer support for group decision making is explicitly considered by *De Hoog, Breuker* and *Van Dijk* (Amsterdam). After a short overview of group decision theory these authors discuss various group decision support (GDS) methods and programs. They particularly discuss their own Program for Assisting Negotiations In Collectivities (PANIC) which incorporates a Multi-attribute Individual Decision Assistance System (MIDAS) based on multi-attribute utility theory. De Hoog et al. point out that vis-à-vis LST decision making, formal analysis methods and programs can only be modest tools. Formal decision supporters always face the dilemma that wishing to improve decision making requires putting a structure on the process, but decision makers may not like this that much, because they desire to maintain personal flexibility and autonomy ("the Scylla is too much structuring; the Charybdis is too much individual freedom").

The next few chapters of the book more directly deal with policy decision making about LST projects. *Hisschemöller* and *Midden* (Rotterdam, Leiden) discuss four different general approaches towards understanding technological risk and dealing with public responses to LST projects. The technical, market, public participation and distributive justice approaches - somehow escalating in 'elaborateness' - each have their advantages and their drawbacks. However, the technical and market approaches seem to be natural allies, as do the public participation and social justice approaches. The authors discuss several case studies from different countries and they conclude, among other things, that three factors seem to determine the success or failure of a government siting strategy: the public perception of risks, the administrative context of policy making, and trust in the responsible authorities.

In the next two chapters two Dutch case studies of LST siting problems are presented: one concerning a noncontroversial 10 Megawatt wind-turbine park, the other involving a highly controversial 'long-term interim' (50-100 years) surface storage facility for radioactive waste from hospitals, laboratories and nuclear power plants. Both cases - one solved, one as yet unsolved - occurred in the wake of the Dutch "Wide Societal Discussion on (nuclear) Energy Policy" (Jansen, 1985; Vlek, 1986), revealing a very positive public opinion about wind energy and a rather negative attitude towards nuclear power.

Lubbers (Arnhem) presents the windfarm case, showing that a multi-criteria (or multi-attribute utility) analysis (MCA) may be performed with inputs from different agencies and with sets of importance weights reflecting different viewpoints ("Poltergeist", "Money", "Research-is-illuminating" and "Money-is-not-important"). The MCA in this case led to a provisionally selected site - near the village of Sexbierum in the province of Friesland - which was then further subjected to various legal and administrative procedures that were to yield the required licences. Lubbers' exposition nicely illustrates one way of tackling both the individual judgment problem and the social interaction and aggregation problem distinguished above.

Brouwer (The Hague) tells the story of Holland's radioactive waste policy since - under pressures of environmental protection groups - it was decided to stop ocean-dumping of low- and middle-level radioactive waste in 1982. A long-term interim storage facility is badly required - in the hope that a definite underground store will become available in the next century. Briefly told: there was a provisional site selection committee paying attention to 'political and administrative suitability' only. Then there were two successive environmental impact analyses, one specifically aimed at the "selected" site. Then followed a delayed outburst of protests from the local population. And now, in the Fall of 1988, a new site, close to the originally selected one but psychologically more acceptable, is being evaluated. One lesson is that controversial politics is not easily captured by formal decision making procedures. Another lesson could be that an organized social decision procedure based on open information and various expertise might have prevented a lot of tension, frustration and misgivings, and might have saved a lot of time, money and paperwork. In an editorial postscript one of us sketches the newest developments around this case.

Almost as if he were looking back on the nuclear energy debate and the radioactive waste storage controversy, *Van der Pligt* (Amsterdam) argues that public attitudes towards radioactive waste are part of a more general attitude towards nuclear energy which generally differs from what the (official) experts think. Opposition to nuclear electric power generation may be considered to stem from an "irrational phobia" disregarding relevant economic, ecological and political arguments; or it may be explained on the assumption that rational, but deviating arguments and evaluations determine public opinion. Research on public risk perceptions has revealed that qualitative dimensions of risk, such as catastrophe potential, controllability and familiarity, are important. Attitude research has shown that different dimensions of the nuclear power issue are differently salient for different interest groups. Van der Pligt suggests that - to resolve public acceptance problems - the quality of the social decision making procedure may be more important than the particular combination of, and trade-offs among estimated risks, costs and benefits.

Public information and communication issues in connection with LST are discussed by *Cvetkovich, Vlek* and *Earle* (Bellingham, Groningen, Bellingham). How could this be promoted, as, e.g., requested by the Commission of the European Communities in article 8 of their 'post-Seveso Directive' of 1982? After delineating the basic components of effective communication the authors successively review the Factual Information Model, the Personal Gain-Loss Model and the Values Model (emphasizing fairness) for designing public information programs concerning hazardous technology. Having discussed the shortcomings of these models, Cvetkovich et al. then present a decision-theoretic model which utilizes elements from both stress-and-perceived-control theory (e.g., Coyne and Lazarus, 1980) and rational decision analysis, to conclude that effective hazard information should enable one to make good decisions among controllable courses of action which do protect one against accidents and/or provide relief in case of disaster.

The special conditions of technological emergency decision making are discussed by *Rosenthal* and *Van Duin* (Leiden). They state that crisis decision making *a fortiori* escapes the rational-synoptic view of Bayesian decision theory: a crisis puts everyday sociopolitical reality into a 'pressure-cooker', thus posing serious threats to basic structures and/or fundamental values of society. One reason for society's neglect of adequate disaster management planning is that the term "technological emergency" seems to be a *contradictio in terminis*: technology being man's highest achievement, it should not run out of control. Based on studies of various different political emergencies, Rosenthal and Van Duin characterize crisis decision making in six propositions: decision making becomes increasingly centralized, 'bureaupolitics' flourish, formal rules and procedures are de-emphasized, decision makers tend to rely on 'trusted, liked sources', information gaps are filled by drawing analogies with previous similar cases ("learning by precedent"), and decision makers are easy victims of 'groupthink' (Janis, 1972). The conclusion is that crisis decision making is seriously deficient compared to 'normal' bureau-political decision making ("muddling through with some satisficing at best"). The recommendation is that crisis decision making should be reviewed and improved on the basis of social decision quality criteria.

The final chapter of the book, by *Vlek* and *Cvetkovich*, distills and relates various general issues from the earlier chapters. We provide a summary characterization of technological decision problems and emphasize the limitations of normative theory for optimally dealing with them. Risk analysis and risk management concepts and methods are discussed, starting from a summary table of formal definitions and cognitive dimensions of risk; the importance of "controllability analysis" is pointed out. The definition and role of experts are briefly considered, as are the confusing public effects of disagreements among experts. We then come back to the distinction between individual judgment and social interaction and aggregation, emphasizing that a clear, well organized decision procedure might help both the individual participant to order his/her thoughts and overcome cognitive limitations, and the group or organization to interact effectively, develop mutual trust and bring its group dynamics reasonably under control.

There are dilemmas here: The dilemma of forced process structuring as against personal autonomy, the dilemma (actually a *tri*lemma) among collective rationality, equal participation and decisiveness in designing or selecting a social decision procedure, and - not the least - the dilemma of participatory (democratic) decision making as against expert-guided, competent decision making.

We dwell a bit more upon computer programs for decision aiding which seem indispensable for handling, summarizing and feeding back the amounts of data that may go around during some kind of organized social decision procedure. And we discuss the need for increased public involvement for which we present four distinct arguments. Finally, we outline a project-oriented social decision procedure comprising ten steps or phases. Meant to be a 'best conglomerate' of various principles, methods and suggestions discussed in earlier chapters, this procedure needs empirical testing and evaluation. The latter applies to several more restricted approaches as well.

Perhaps one might say that there is a challenge here: for decision theorists to undertake some serious field research, for research-promotion organizations to fund proposals that may look unusually ambitious, and for political authorities (the real problem owners) to make LST problems available for, and to encourage relevant parties to get involved in, realistic try-outs of a social decision procedure that might actually support effective policy making concerning technological developments.

REFERENCES

Abrams, N.E. (1979). Nuclear politics in Sweden. *Environment, 22(4),* 6-40.

Coyne, J.S. and Lazarus, R.S. (1980). Cognitive style, stress perception and coping. In: I.L. Kutash and L.B. Schlesinger (Eds.), *Handbook on Stress and Anxiety* (pp. 144-158). San Francisco: Jossey-Bass.

Gamble, D.J. (1978). The Berger Inquiry: an impact assessment process. *Science, 199 (March),* 946-952.

Gregory, R. and Lichtenstein, S. (1987). A review of the high-level nuclear waste repository siting analysis. *Risk Analysis, 7(2),* 219-224.

Janis, I.L. (1972). *Victims of groupthink.* Boston: Houghton-Mifflin.

Jansen, J.L.A. (1985). Handling a debate on a source of severe tension. In: E. Denig and A. van der Meiden (Eds.), *A geography of public relations trends.* Dordrecht: Martinus Nijhoff.

Nelkin, D. and Pollak, M. (1978). The politics of participation in the nuclear debate in Sweden, the Netherlands and Austria. *Public*

Policy, *25(3)*, 333-357.

O'Riordan, T., Kemp, R. and Purdue, M. (1985). How the Sizewell Inquiry is grappling with the concept of acceptable risk. *Journal of Environmental Psychology, 5*, 69-85.

Parker, Mr. Justice (1978). *The Windscale Inquiry*. London: Secretary of State for the Environment.

Perrow, Ch. (1984). *Normal accidents; living with high-risk technologies*. New York: Basic Books.

Renn, O., Stegelmann, H.U., Albrecht, G. and Kotte, U. (1984). The empirical investigation of citizens' preferences with respect to four energy scenarios. *Technological Forecasting and Social Change, 26*, 11-46.

Royal Commission on Electric Power Planning (1980). *Final Report*. Toronto, Ontario: Queens Printer.

Thomas, K.W. (1976). Conflict and conflict management. In: M. Dunnette (Ed.), *Handbook of Industrial and Organizational Psychology*. Chicago: Rand McNally.

Vlek, C.A.J. (1986). Rise, decline and aftermath of the Dutch 'Societal Discussion on (nuclear) Energy Policy' (1981-1983). In: H.A. Becker and A. Porter (Eds.), *Impact Assessment Today* (pp. 141-188). Utrecht (Neth.): Van Arkel.

Wynne, B. (1982). *Rationality and ritual: the Windscale Inquiry and nuclear decisions in Britain*. Chalfont St. Giles (Bucks.): British Society for the History of Science.

SOCIAL CHOICE THEORY AND PRACTICE

THOM BEZEMBINDER

Nijmegen Institute for Cognition Research and Information Technology (NICI)
University of Nijmegen

1. INTRODUCTION

With Coombs and Avrunin (1977a, p. 226) we may distinguish three kinds of conflict:
' (a) conflict within an individual that arises because he is torn between incompatible goals
  (b) conflict between parties that arises because they want different things and must accept the same thing ...;
  (c) conflict between parties that arises because they want the same thing and must accept different things ...'.

The area of social choice (SC) as traditionally conceived deals with a community making a decision starting from and taking into account the preferences of its members. The preferences of the individuals being diverse and the decision sought by the community being unique, SC constitutes a conflict of kind (b). The decision of the community should be based on and in some sense reflect the preferences of the individuals. How to realize this maxim is the concern of SC theory.

Coombs and Avrunin proposed the above distinctions only for those kinds of conflict in which the alternatives the conflict is about are clearly distinguished and given a priori. So, the parties in the conflict do agree about what the alternatives are and they know that eventually the conflict is being resolved in one of the pre-given alternatives. Considering a set of cars ordered from low to high quality and, jointly, from low to high price - so that higher quality can only be obtained at higher price - a prospective buyer is torn between low price and high quality (conflict of kind (a)) but eventually decides on one of the pre-given cars in which (s)he deems the trade-off between price and quality optimal. In deciding where to store nuclear waste first a set of feasible sites is determined and, then, one of those sites is eventually chosen although parts of the community may have various and strong opinions on what site is best

15

Ch. Vlek and G. Cvetkovich (eds.), Social Decision Methodology for Technological Projects, 15–37.
© *1989 by Kluwer Academic Publishers.*

or worst (conflict of kind (b)). Two candidates applying for the same job are involved in a conflict of kind (c). National elections serve for determining which of the pre-given political parties do and which do not come to power although power is all parties' wish (conflict of kind (c)).

In quite a few conflicts the alternatives the conflict is about are neither clearly distinguished nor given a priori. Conflicts between management and labor are rarely confined to wages. In that case the alternatives of the conflict would be given a priori and the conflict is of kind (b). Usually, however, negotiations between management and labor are about portfolios containing a diversity involving wages, working conditions, safety regulations, (un)employment policy, sickness premiums, side-payments, and the like. Deliberations about building a nuclear power plant are highly complex and deal with economic interests, safety warrants, waste storage, environmental interests, long trend energy analyses. These kinds of conflict are eventually solved in an alternative not given a priori but constructed in the very process of decision making.

In its present state SC theory deals mainly if not exclusively with conflicts about pre-given and well-distinguished alternatives. This part of SC theory originated from welfare economics and has a vast literature in which two topics are particularly important: (i) the impossibility of constructing fair collective choice rules due to inconsistencies in the requirements desired for a fair rule - see Kelly's 1978 review, and (ii) the (im)possibility and nature of interpersonal comparisons of utility - see Möller's 1983 review. As yet, hardly any SC theory has been developed on conflicts about ill-defined or not pre-given alternatives. In the sequel we will first consider conflicts about pre-given alternatives and, then, conflicts about not pre-given alternatives.

2. THE CASE OF PRE-GIVEN WELL-DEFINED ALTERNATIVES

In the classical form of the SC problem we have for each member i of a community N an individual weak order of preference R_i on a given set Z of alternatives and we try to establish a social weak order of preference R on Z, that is, we look for a *social welfare function* (SWF) - see Appendix. The required SWF, however, should be a fair aggregate of the individual preferences. The SWF pre-eminently well-known is the *rule of majority decision* (RMD). Letting $M(xy)$

denote the number of people for whom xR_iy holds, RMD says that for all $x,y \in Z$

$$M(xy) \geq M(yx) \implies xRy . \tag{1}$$

In Western societies RMD is generally accepted and by and large felt to be fair. Yet, it has at least two drawbacks. First, it does not always lead to a transitive outcome. For instance, the individual preference orderings *xyz, yzx, zxy* result into the cyclical majorities $M(xy) = M(yz) = M(zx) = 2$ while $M(yx) = M(zy) = M(xz) = 1$ so that xRy, yRz, zRx by (1) and, hence, in this case R is not transitive. This phenomenon, first noted by Condorcet (1785) and known as *the voting paradox*, is commonly felt to tarnish the fairness of RMD since it opens ways for manipulation. Indeed, with cyclical majorities using a sequence of pairwise knock-out votings the outcome depends on the order of the pairwise decisions. To be sure, if in the above example x and y are compared first, then y is knocked out and z wins in the remaining pair (x,z) whereas x wins if y and z are compared first. So, if the social choice is determined by sequential knock-out votings, shrewdly advancing a particular order of pairwise decisions may enhance one's prospects of getting one's way. More generally, the majority rule has strategy aspects and, in particular, misrepresenting once's true preferences may sometimes help to enhance one's interests (Luce and Raiffa, 1957, p. 359; Sen, 1970, p. 193). - An other drawback of RMD is that it does not take into account intensities of preference. All votes are counted equal irrespective of whether they are based on weak preference hardly distinct from indifference or on intense preference stemming from delight or disgust. It may be deemed unfair that the halfhearted and the ardent have equal influence. Taking intensities of preference into account requires that the utilities of the alternatives for the individuals are compared between individuals. It is a contested issue whether or not such comparisons are possible and, if so, how they may be obtained. Be this as it may, SC theory considers two classes of procedures of social choice, namely, procedures that do and do not involve such comparisons.

2.1 Procedures without interpersonal utility comparisons

Let F be a rule for amalgamating a set of individual preferences R_i into a social preference relation R (see Appendix). The affective connotations of fairness set aside, in SC theory fairness is commonly defined in terms of a set of constraints on F. So, for considering a rule of amalgamation as fair we may require that it leads to transitive outcomes and/or takes into account intensities of preference. We may also require that the outcome does not in any way depend on the persons expressing the preferences so that the outcome stays fixed when all R_i are preserved but expressed by different persons. Other constraints on F putatively characterizing a fair SWF come easily to mind: fairness is a human concept for which there are no observations in nature.

However, also on fairness it is easy to ask too much. This was first demonstrated dramatically by Arrow's (1951) celebrated impossibility theorem in which a set of seemingly innocuous constraints for an SWF is shown to be inconsistent. Arrow studied the problem of finding a 'fair' SWF, say F, for aggregating a set of n individual weak preference orders $R_1,...,R_n$ - all defined on a common set Z of alternatives - into a social weak preference order R so that

$$F(R_1,...,R_n) = R \qquad\qquad (2)$$

with 'fairness' being specified by a set of constraints on F and R. As an attempt to escape the voting paradox R was required transitive. This requirement is also called the requirement of *independence of path* since it is intended to keep off the order of pairwise decisions from determining the final outcome. Furthermore, R was required *decisive* in all pairs (x,y), that is, complete - see Appendix.

Arrow's constraints on F are:
(i) *Unrestricted Domain*: F should result into a social weak order of preference for any n-tuple of individual weak preference orders;
(ii) *Weak Pareto Principle*: If for any x,y all n persons strictly prefer x to y then in the social preference order x should likewise be strictly preferred to y;
(iii) *Independence of Irrelevant Alternatives* (IIA): For any x,y the social preference on x and y only depends on the individual preferences on x and y and does not depend on the individual preferences on any other alternatives. Blair and Pollak (1983) call this constraint 'Pairwise Determination' which is a much more appealing term. Yet, we stick to Arrow's term since the abbreviation IIA is quite current;
(iv) *Nondictatorship*: there is no person such that, for any x,y, whenever (s)he strictly prefers x to y then in the social preference x is also strictly preferred to y.

Arrow showed that for three or more alternatives and a finite number of at least two persons any SWF satisfying (i), (ii), and (iii) is dictatorial in the sense that it violates (iv), that is, there is a dictator as specified in (iv). So, there does not exist an SWF that satisfies all Arrow's constraints. Attempts to weaken the constraints proposed by Arrow have generally led to quite a variety of other - often equally fascinating - impossibility theorems on social welfare as well as social decision functions - see Appendix. For reviews see Sen (1977b) and Kelly (1978).

We suffice to elaborate on IIA which in the literature on Arrow's constraints gets particular attention. The notion of IIA is perhaps most easily clarified by the example given by Luce and Raiffa (1957, p. 341). Consider the persons i and j who say to evaluate the alternatives x and y amongst the following prizes and fines as follows from least preferred to most:

person i :	x	-1000	-1	0	1	1000	y
person j :	-1000	-1	y	0	x	1	1000

Now, if i and j are to determine a common preference on x and y it seems intuitively plausible that they should prefer y to x. This conclusion, however, is prohibited by IIA which forbids to take into account any alternatives different from x and y. Indeed, those other alternatives may actually not even be feasible and, so, an SWF not obeying IIA may tempt individuals to strategically misrepresent their preferences. For instance, it may very well be that person i can hardly discriminate his/her true preferences for y and prize 1 but in order to avoid x to be first in the common preference, it behooves i to express the evaluation of y as exceeding prize 1000. This kind of misrepresentation is enervated by IIA. Making manipulation useless and preventing SC from becoming a game of strategy are important motives to impose IIA as a constraint on a fair SWF. A fair rule should not favor the shrewd.

The problem of fairness in SC is like the problem of rationality in multi-attribute individual choice which deals with conflicts of kind (a). Both fairness and rationality are human concepts, and delineating their contents are problems of human agreement. In SC theory rationality has commonly been taken as transitivity and, particularly, as complying with the requirement of path independence. According to Fishburn (1970), however, the irrationality of transitivity is equally tenable.

Consider 100 individuals $i_1,...,i_{100}$ with the following weak preference orders on the 101 alternatives $x_1,...,x_{101}$:

i_1	:	x_1	x_2	x_3	...	x_{98}	x_{99}	x_{100}	x_{101}
i_2	:	x_2	x_3	x_4	...	x_{99}	x_{100}	x_{101}	x_1
	:								
i_{100}	:	x_{100}	x_{101}	x_1	x_2	...	...	x_{98}	x_{99}

Let the 100 individuals determine their pairwise social choices in accord with

$$\text{for any } x,y : M(xy) \geq 99 \implies xPy \qquad (3)$$

where $M(xy)$ is to be read as in (1). Now, except i_1 all individuals prefer x_{101} to x_1, except i_2 all individuals prefer x_1 to x_2, ..., except i_{100} all individuals prefer x_{99} to x_{100}. So, $M(x_{101}x_1) = M(x_1x_2) = ... = M(x_{99}x_{100}) = 99$ so that $x_{101}Px_1, x_1Px_2, ..., x_{99}Px_{100}$. Hence, when requiring (3) to be transitive we obtain $x_{101}Px_{100}$. However, *all* 100 individuals prefer x_{100} to x_{101}! Is transitivity a rational requirement?

Let us now generalize (3) to:

$$\text{for any } x,y : M(xy) \geq n\text{-}1 \implies xPy \qquad (n \geq 2) \qquad (4)$$

and let us consider the first two preference orders of the example showing the voting paradox in Section 2, namely, i_1:xyz and i_2:yzx. Except i_1 all individuals prefer z to x while, except i_2, all individuals prefer x to y. Hence, using (4) for $n = 2$ we have zPx and xPy. Requiring transitivity we obtain zPy but all two individuals prefer y to z. So, Fishburn's example does not add anything new to the voting paradox but it does show its dramatic impact.

The social preference order sought by Arrow is a weak order R on Z. Since I (indifference) and P (strict preference) unite into R so that $R = P \cup I$ (see Appendix), the transitivity of R may be split up into transitivity requirements on P and I. So, we may require transitivity on either P or I or both. For a given pair of alternatives (x,y) consider as SWF the *consensus rule*

$$xR_iy \text{ for all } i \in N \text{ \& } xP_iy \text{ for some } i \in N \implies xPy .$$

Sen (1970, p. 52) showed that this SWF satisfies all Arrow's requirements except for the resulting social ordering being only *P*-transitive and not *I*-transitive. However charming this result, it is obtained at the cost that we only reach decisions in exceptional cases, namely, only in those pairs (x,y) in which all individuals agree.

Whereas the consensus rule requires unanimity of all *n* individuals, Fishburn's rule (4) requires unanimity of any *n*-1 individuals. So, Fishburn's rule will generally be (somewhat) more decisive than the consensus rule but it does no longer necessarily lead to a transitive result. More generally, we may consider *sub-consensus rules* requiring unanimity of any *n-t* individuals with $t < n$. For *t* increasing from 0 to *n*-1 the resulting rules will generally be more decisive - in the sense that they will lead to a social decision in more pairs of alternatives - but, at the same time, they will allow social decisions that, in general, are increasingly nontransitive. Setting aside the complex question in what sense structures of binary choices may be said to be more or less nontransitive (Bezembinder, 1981), so much is clear that, as pointed out by Blair and Pollak (1983), decisiveness has to be paid for by rationality in the sense of transitivity.

The sub-consensus rules are egalitarian in the sense that, in a given pair (x,y), no matter which *n-t* individuals show consensus, they determine the social preference *P* in (x,y). We may also consider a rule of SC - denoted rule D1 - that designates a particular individual as a dictator with the power as specified in the constraint of Nondictatorship. Rule D1 would of course violate that constraint but, in addition, it would be nonegalitarian since it designates a specific individual - not just anyone - as a dictator. Clearly, rule D1 is maximally decisive since it follows the dictator's preferences in all pairs (x,y). Next, we may consider a rule D2 in which an oligarchy of two individuals is designated as a dictator in all pairs (x,y) in which these two individuals are unanimous. In general, rule D2 will be less decisive than D1 but it allocates power to 2 persons instead of a single one. Like D1 rule D2 is also nonegalitarian since it designates two particular individuals as the oligarchy rather than any two individuals. More generally, we may consider the nonegalitarian rules D2,...,D*n* designating, respectively, oligarchies of 2,...,*n* persons as a dictator. The oligarchies designated in D1,...,D*n* will, respectively, comprise increasingly more individuals but, going along the way, they will be decreasingly decisive. Actually, D*n* coincides with the consensus rule!

Indeed, as shown in the beautiful essay by Blair and Pollak (1983), asking for a social choice rule that is egalitarian, decisive and rational (in the sense of transitive) is asking too much. In constructing such rules more decisiveness has to be paid for by sacrificing more of either rationality (in an egalitarian rule) or equality of power (in a nonegalitarian rule). More generally, as pointed out by Richelson (1975, 1978) for extensions of RMD, all voting rules have their own assumptions. So, there is no lack of rules but every rule has its price.

2.2 Procedures involving interpersonal utility comparisons

Letting $u_i(x)$ denote the utility of alternative x for individual i we consider the following SWFs:

$$\sum_{i \in N} u_i(x) \geq \sum_{i \in N} u_i(y) \implies xRy ,\tag{5}$$

$$\underset{i \in N}{\text{minimum}} [u_i(x)] \geq \underset{i \in N}{\text{minimum}} [u_i(y)] \implies xRy .\tag{6}$$

Rule (5) is the utilitarian SWF which chooses the alternative with the maximal total utility - utilities being added over individuals. Rule (6) is the maximin rule that maximizes the welfare of the worst-off individual. This rule is often attributed to Rawls (1971) except - as pointed out by Sen (1980; 1982, p. 359) - by Rawls himself! Either one of these rules is only meaningful provided the inequality in the condition is preserved under all permitted transformations of all individual utilities. Considering (5), let there be an individual j for whom $u_j(y) > u_j(x)$. Incidentally, this is a rather weak assumption that only prevents (5) to be trivial. Now, if u_j is an ordinal scale we are free to increase $u_j(y)$ to, say, $u_j^*(y) > u_j(y)$. However, when only transforming u_j - and keeping all other u_i with $i \neq j$ fixed - we may easily choose $u_j^*(y)$ such that the inequality in (5) is reversed. So, in order to preserve that inequality the scales u_i for all $i \in N$ may clearly not be transformed independently. Actually, it is easily seen that for preserving the inequality in (5) it suffices that all u_i are interval scales with a common unit, so-called *unit comparable interval scales*. True, this condition is not necessary for preserving the inequality but it is not apparent how it may be weakened in a simple way. Similarly, for preserving the inequality in (6) it suffices that the u_i satisfy *ordinal level comparability* in which case there exists a positive monotone transformation Φ such that $u_i^* = \Phi(u_i)$ for all $i \in N$. To be sure, the conditions of unit comparable interval scales and ordinal level comparability both preclude the u_i from being transformed independently and only allow that the u_i are transformed conjointly. Comparability constraints are interpersonal constraints: they tie the u_i for all $i \in N$ together in the sense that the u_i may only conjointly be transformed.

It is a well-known result in the theory of measurement (Krantz, Luce, Suppes and Tversky, 1971, p. 15, theorem 1) that a weak order may be represented by an ordinal scale. In the amalgamation (2) considered by Arrow (1951) both the R_i for all $i \in N$ as well as R are weak orders. Hence, we may write (2) in the form

$$G(u_1,...,u_n) = u \tag{7}$$

which says that a procedure G amalgamates the individual ordinal utility scales u_i for all $i \in N$ into a social ordinal utility scale u. Now, since in Arrow's problem the R_i in (2) are supposed to be independent, the u_i in (7) are also independent. The set of independent R_i in (2) is said to be the *informational basis* for the SWF sought by Arrow, that is, the rule of amalgamation F in (2). Sen (1977a) considered to enrich this informational basis by imposing comparability constraints on the u_i in (7). As regards (5) and (6), which are special cases of (7), Sen (1977a) showed the following. (i) Provided the u_i in (5) are unit comparable interval scales, rule (5) satisfies all Arrow's constraints. (ii) Provided the u_i in (6) satisfy ordinal level comparability, rule (6) satisfies all Arrow's constraints except for the Weak Pareto Principle being replaced by the

Strong Pareto Principle: For any x,y if xR_iy for all $i \in N$ then xRy while if, in addition, xP_iy for some i then xPy.

Clearly, provided the u_i satisfy appropriate comparability constraints, rule (5) meets all Arrow's requirements while the same holds for (6) with a slight modification of those requirements. So, we now have possibility theorems instead of an impossibility result. More generally, by imposing other comparability constraints on Arrow's informational basis Sen (1977a) obtained additional possibility theorems. In this sense, obtaining fair rules of SC hinges on the (im)possibility of interpersonal comparisons of utility.

In spite of the vast literature on interpersonal comparisons of utility - see Möller's 1983 review - their (im)possibility is still contested and their nature elusive. Also, it is not clear how they may be obtained if only for lack of a well-established definition. Even Sen who claims that these comparisons can be made (Sen, 1985, p. 52), concedes that '... practical interpersonal comparisons of utility are not particularly easy to make anyway' (Sen, 1985, p. 53). Furthermore, '... the problems of intelligibility and practicality of interpersonal comparisons are not easy to resolve' (Sen, 1986, 221-2). We suffice with one distinction on the source of these comparisons. First, the social preference R in (2) and the social utility u in (7) may be conceived of as to be assigned to an outside observer or an outside planner who, to that end, compares the individual utilities. This outsider then faces the question of how to arrive at these comparisons. Is (s)he to estimate the individuals' utilities? Should (s)he (also) go by behavioral cues? In this view, the problem of these comparisons becomes part of the even more elusive problem of the (im)possiblility of knowing other minds. Second, we may consider to measure the u_i by conjointly constraining the individuals' behaviors that in view of the measurements have to be observed. This approach was suggested by Narens and Luce (1983) who, to the best of our knowledge, are the first authors who proposed a formal definition of interpersonal utility comparisons. However, as these authors themselves admit, the constraints they proposed for the individuals' behaviors do not seem very realistic.

3. THE CASE OF NOT PRE-GIVEN ALTERNATIVES

In complex social decision problems the options to be decided upon are generally not given a priori. Instead, these problems are mostly resolved by a process of deliberations and negotiations directed at creating an option that all parties finally accept after a process of give and take. In general, the option sought for is a portfolio comprising a compromise of a diversity of interests. The negotiations serve to clarify the decision problem, delineate its constraints, detect and clarify the aspects of evaluation for the parties involved, bring about trade-offs and concessions, estimate and fairly distribute gains and losses, and to attend to fair procedures.

Apart from structuring the decision problem deliberations also bring about changes in the people. In the participating people and interest groups deliberations may foster feelings of being a community facing a common problem and sharing common responsibilities. Research on moderately sized groups reported by Wilke (this volume) indicates that in social dilemmas where a public good has to be paid for by sacrificing on self-interest, cooperative behavior is not uncommon. The hypothesis that 'everyone should make a defective choice while leaving it to others to provide the public good, must be rejected in favor of a more moderate version'. Furthermore, 'communicating groups are more cooperative than non-communicating groups' (Wilke, this volume). So, the very process of deliberations may apparently transform individual preferences from being self-centered to being more community-directed.

Being normative, SC theory is to recommend actions. Yet, in its current form SC theory can not recommend deliberations or negotiations directed at creating an outcome not known before, no matter how effective such procedures may be. We mention two reasons for this. First, SC theory only considers pre-given well-defined alternatives and preferences pertaining to such alternatives. Accordingly, preferences are not supposed to change during a decision process - not even to change to be more community-directed. Since, strictly, amalgamations as such are instantaneous it is actually doubtful whether decision processes are at all in the domain of SC theory. Second, the alternatives SC theory considers for amalgamation are potential solutions for the problem at hand so that, in particular, decision procedures which are not solutions are being excluded. However, people do have preferences for the process by which a decision is obtained. Irrespective of how fair a mechanism of social choice may be and no matter what axioms of fairness it may satisfy, people may nevertheless prefer an other procedure, especially a procedure in which they may contribute to shaping the alternatives.

Decision problems in which the alternatives to be decided upon are not pre-given, are ill-defined in the sense that there exists no alternative which, when offered as a solution will be recognized as a solution. Rather, such a problem needs to be clarified while a solution is to be created in a process of consultations and deliberations. For helping this process along several procedures may be used such as cost-effectiveness and cost-benefit analysis, mathematical optimization, and multi-attribute utility measurement (examples of the use of these procedures in complex decision problems are in Bell, Keeney and Raiffa 1977). The 'value oriented social decision analysis' proposed by Chen and Mathes (this volume) uses these procedures but also surpasses them in the sense that it is especially directed to 'increase communication among multiple interests groups ...' and to 'enhance the communication processes that precede the actual decision making stage' (Chen and Mathes, this volume). Chen and Mathes' procedure guides the process of deliberations by structuring the decision problem at hand into several components and using each component for advancing clarity and trust.

For dealing with such deliberations fostering a vision for the social good and a sharing of responsibility Phillips (1984; see also Phillips, this volume) considered *requisite modeling*. Phillips' essay is lucid as regards the practical aspects of such deliberations. It presents clear recommendations on the structuring of the deliberations, the participants, the issues to be dealt with, procedures, and even ways for enervating manipulation. The process of consultations and deliberations is to result into a new social reality characterized by a common understanding of the problem and an idea on what to do. By confronting the different views of the problem, exploring its complexities and resolving inconsistenties, the participants' initial feelings of unease gradually diminish in the course of the deliberations. Then, when no new intuitions arise any more and the participants have an idea of what to do next, a *requisite model* is said to have been developed.

Facing a complex decision, a common understanding of the problem and an idea of what to do are, in general, about as much as may reasonably be desired. Yet, however clear Phillips' practical recommendations may be, their conceptual underpinnings are not any more transparant than an ill-defined problem. The idea of a requisite model resides in the shady area between descriptive and normative theory. A requisite model neither describes what people do nor prescribes what they should do. 'The process of building a requisite model is sometimes conducted in a group, at other times by a succession of discussions between specialists and individual problem owners. But in all cases, the process is consultative and iterative ...' (Phillips, 1984, p. 36). A requisite model is a dynamic concept. Like a work of art its creation requires a process, its completion a decision. And, when finally decided upon, it is recognized while never seen before.

Despite Phillips' plausible procedure, not even if all participants are induced to community-centered preferences do deliberations of a moderately sized group necessarily lead to an outcome that all participants accept. All decision problems have a time limit and for a variety of reasons a common understanding may not be realizable within that limit. Also, because of dissensions on value, reaching unanimity may be virtually impossible. So, eventually it may be unavoidable to use a formal procedure for obtaining a social choice by amalgamating the individual preferences on the alternatives created in the course of the deliberations. The problem is then by necessity reduced to a decision on pre-given alternatives.

4. THE POSITION OF POWER IN SOCIAL CHOICE

In social decision making power is a ubiquitous phenomenon on which there is a vast literature. We suffice with a few remarks.

In complex decision problems the alternatives to be decided upon are usually created in a process of consultations and deliberations. Now, however fair a community's rules of social choice are felt to be and however community-centered its members' preferences, power is inevitably to play a role at two points. First, power plays a role in the process of deliberations if only since people differ in motivation, persuasiveness and gullibility. Second, if by whatever (fair) procedure a final decision has been made, power may very well be needed to keep it stick. Considering the four bases of power mentioned by Wilke (this volume), namely, coercion, compensation, information and legitimation, it is obvious that fair rules of social choice resulting from SC theory are to be executed and administered by legitimate power.

If a solution for a complex decision problem has been created in a process of deliberations, decided upon via a rule of SC generally felt to be fair and, then, maintained or, if necessary, enforced by a legitimate power, it is clearly not only power by which the problem has been solved. So, Wilke's (this volume) view that 'public goods are provided and maintained by power' is perhaps a bit one-sided. According to Wilke, power is especially important in solving social dilemmas that come into being because 'any participant is more attracted to take the advantage of the provision ... and letting others incur the disadvantages ...' (this volume). Yet, the experimental results Wilke brings forward are more differentiated in that they suggest that voluntary sacrifices for social goods are not uncommon. However, these results pertain to moderately sized groups whereas social dilemmas may very well be at a grand, say national, scale. But

even then, whether, for instance, taxes are only maintained by power
is a moot question.

5. TOWARDS PRACTICAL PROCEDURES OF SOCIAL CHOICE

A striking result of SC theory is a variety of impossibility theorems
on both social welfare and social decision functions. This indicates
that in SC theory the concept of fairness constitutes a problem of
optimization rather than definition. To be sure, every impossibility
theorem may be regarded as an abortive attempt to define a fair
procedure of social choice. True enough, every possibility theorem
provides a definition of fairness that we may try to put in operation
but it has been obtained at the price that some constraints seemingly
characterizing a fair rule are not part of the definition. Indeed, since
fairness is a human concept that evolves with humans and society,
there is no natural limit to the constraints a fair rule is to meet.
Such a limit is itself subject to human agreement. With respect to the
constraints for a fair rule compromises are inevitable.

Compromises on the constraints for a fair rule of social choice may
also be obtained by alleviating requirements on the measurement of
individual utilities. Although such measurements do not by themselves
affect fairness the measurement problem is relevant here for two
reasons. First, we want truthful measures proof against manipulation.
Second, SC rules that amalgamate individual preferences into a social
preference may require some form of interpersonal utility comparisons
to the effect that, like in (5) and (6), the individual utilities can not
be measured separately but have to meet a common constraint. So,
with rules of social choice like (5) and (6), the problem of measuring
utilities can not dodge the problem of interpersonal utility
comparisons.

Rules of SC based on compromises as just indicated are easily found.
The sub-consensus rules (see Section 2.1) considered by Blair and
Pollak (1983) rest on trade-offs between decisiveness and either
(non)transitivity or (in)equality of power. Letting 'total utility' denote
a sum of individual utilities Sen (1980; 1982, p. 359) writes: 'In case of
utilities, if they are taken to be observed facts, total utility will not
be counter-factual. Thus total utility equality is a matter for direct
observation ...'. Taking utilities as factual matters accessible for direct
observation - an idea not shunned by Sen - an outside planner may
use (5) or (6) provided (s)he is prepared to make the appropriate

assumptions on interpersonal comparability of utilities. So, practical rules of SC may be obtained by mitigating requirements on fairness or measurement.

It may be argued that such mitigations seem to come down to rather blunt assumptions. As yet, however, SC theory has hardly considered what compromises as meant above it may offer for practical problems or what alleviations of constraints on fairness or measurement may be regarded acceptable. As a consequence, for designing practical procedures at the cost of seemingly acceptable assumptions SC theory can offer very little. Dealing with SWFs and, specifically, considering (5) Bell, Keeney and Raiffa (1977, p. 6) write: '... the literature [on SWFs] theorizes about idealized behavior and little attempt has been made to assess actual utility functions and to use these to guide behavior' ([...] added). Indeed, by considering trade-offs between desirable and undesirable properties of SC rules Blair and Pollak (1983) set an exceptional example worth following.

Practical procedures of SC may also be obtained by structuring the decision problem at hand. In particular, it is advisable to try to construct a set of alternatives on which all individuals' preference functions are single-peaked since in that case - as is well-known (Luce and Raiffa, 1957, p. 356) - both simple majority rule and any weighted majority rule result into a transitive outcome provided the number of persons is odd. Consider a preference function on a linearly ordered set $Z = \{x,y,z,...\}$ so that, say, $x < y < z < ...$ and let R be a weak preference order on Z (see Appendix). Following Aschenbrenner (1981, p. 229) we now define:

> A weak preference order R on Z is *single-peaked* with respect to
> $<$ if and only if for all $x < y < z \in Z$
> (i) $xPy \implies yPz$ & (ii) $zPy \implies yPx$.

So, in any $x < y < z \in Z$ single-peakedness is violated if the intermediate alternative y is least preferred. Figure 1 shows a set of single-peaked preference functions of 3 persons on a common linearly ordered set of alternatives $\{x,y,z,w\}$ with $x < y < z < w$. Note that the pairwise majority decisions on the 6 pairs of alternatives are transitive and result into the social preference order $yRzRwRx$. However, if the preference function of person k is replaced by the dashed preference function of person k^* in which the intermediate alternative y is least preferred, then we obtain the nontransitive pairwise majorities *xRy, yRz, zRw,* and *wRx.*

Figure 1: Three single-peaked preference functions and one single-dipped preference function on four objects.

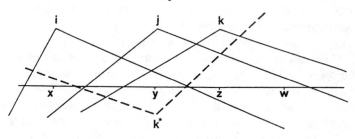

Individuals: {*i, j, k, k**}; Alternatives: {*x, y, z, w*}
Preference orderings   *i* : *xRyRzRw*
                       j : *yRzRwRx*
                       k : *zRwRyRx*
                       *k**: *wRzRxRy*
Majority orderings   (*i, j, k*) : *yRzRwRx* (transitive)
                      (*i, j, k**): *xRyRzRwRx* (non transitive)

In accord with Williamson and Sargent (1967), Sen (1969) points out that the requirement of the odd number of persons is not to be taken lightly. Specifically, it is required that 'the number of non-indifferent individuals expressing themselves with respect to every triple be odd ...' (Williamson and Sargent as quoted by Sen (1969; 1982, p. 129-130). Yet, if we try to construct a choice function (see Appendix) instead of a social ordering, then 'the restriction on the number of individuals can be dispensed with' (Sen 1969; 1982, p. 130). Also, when trying to construct a SWF, changing an individual indifference into a strict preference does not seem insuperably difficult unless the indifference results from repugnance to both alternatives as in some moral dilemmas.

Coombs and Avrunin (1977a, b) developed a theory to account for the occurrence of single-peaked preference functions and they showed that this theory 'has some bearing on the logic of conflict and its resolution' (1977a, p. 218). We will not go into this theory but suffice with a few general remarks directed at a result that is useful for obtaining a practical procedure of social choice. Coombs and Avrunin consider options of which the attractive and unattractive aspects are indissolubly interwoven (for instance, a more lasting car is more expensive). A set of such options is called an *efficient set* if 'the attractive and unattractive aspects are ordered the same within the options so that the decision maker has to put up with more of each unattractive attribute if he desires more of some attractive attribute. Further, it is required that the unattractive attributes increase relatively faster than the attractive ones' (Aschenbrenner, 1981, p. 229). Coombs and Avrunin showed that efficient sets of alternatives are important as a mediate for obtaining single-peaked preferences.

Given a set of alternatives let us join all attractive attributes into one dimension and all unattractive attributes into an other. We now have a set of two-dimensional alternatives, say, $\{x = (x_1, x_2),\ y = (y_1, y_2),\ ...\}$ in which, say, the first dimension represents the attractive attributes and the second the unattractive ones. When selecting a best alternative from this set it comes as a matter of course to eliminate all *dominated* alternatives, i.e. all alternatives that on both dimensions are less preferred than an other. Specifically, if x dominates any y in the sense that x_1 is preferred to y_1 and x_2 to y_2, then y will no longer be considered and is eliminated. Using this elimination principle we are left with a set of alternatives that may very well be (almost) efficient. In Aschenbrenner's (1981, p. 230) words: 'Eliminating all dominated alternatives from a two-attributed set would yield a set ... coming close to being efficient'. Furthermore, Coombs and Avrunin (1977a, p. 226) write 'Although this [elimination] principle does not imply efficiency, it does imply that options varying on only two dimensions are ordered in the same, or opposite, ways on both dimensions, which is a necessary condition for efficiency' ([...] added).

For instance, consider the two-dimensional alternatives

(9,1), (8,4), (7,2), (6,3), (5,6), (3,5), (2,8),

in which the first and second numbers represent ordinal utilities on the first and second dimension, respectively. Assuming that on both dimensions a higher utility is preferred to a lower one, we notice that (7,2) and (6,3) are dominated by (8,4) while (3,5) is dominated by (5,6). Eliminating the dominated alternatives we obtain

(9,1), (8,4), (5,6), (2,8),

in which the preferences on the two dimensions are opposite. The less attractive the options are on the first dimension (e.g., effect on the ecology or social cost) the more attractive they are on the second (e.g., energy output or social benefit). In certain contexts it may also be appropriate to interpret the two dimensions as reflecting the opposite interests of two (political) parties. Clearly, applying the above elimination principle to a set of two-dimensional alternatives we are left with a set of alternatives in which a trade-off is required between one dimension (party) and the other. Structuring a decision problem in this way gives clarity on the domain of conflict and thus provides a simple specification of the nature of feasible concessions. And what trade-off to choose may even be accessible to public discussion and judgment.

6. THE QUEST FOR MORALITY

Intuitively, generality is at the core of fairness. Imposing a constraint on a rule of social choice precludes a particular form of unfairness and thus establishes fairness in a more general sense. However, as appears from the variety of impossibility theorems in SC theory (see Kelly's 1978 review), it seems that in SC theory we have been aspiring too much of generality. On the other hand, since traditional SC theory only considers amalgamating individual utilities leaving all non-utility information out of consideration, SC theory may very well be not general enough. Amalgamating individual preferences into a social preference is an exercise within the framework of *welfarism* in which a social state is valued exclusively after the utilities in that state. However, should we take individual utilities at face value and accept for amalgamation without inquiring after their origin and nature?

The limitations of welfarism have been convincingly exposed by Sen (1979) both for the case in which the information on the individual utilities is poor and does not exceed the level of noncomparable ordinal scales as well as for the case in which it is rich. Being a normative theory on an ethically relevant subject, SC theory can not but reckon with ethical considerations in questions about how to construct fair rules of social choice as well as what utilities to accept for amalgamation. Consider an individual 'who finds that he enjoys seeing others in positions of lesser liberty ... The pleasure he takes in other's deprivation is wrong in itself ...' (Rawls, 1971, p. 31, as quoted by Sen 1980; 1982, p. 362). So, it is a dubious proposition that all individual utilities, irrespective of their nature and source, may serve as a basis for social choice. Indeed, 'ethical preferences rather than personal likings and dislikings of individuals should constitute the basis of social preference' (Pattanaik, 1971, p. 169). So, when observing individual preferences by word of mouth we should ask 'Do you think that x is a better social state than y?' rather than 'Do you prefer x to y?' (Pattanaik, 1971, p. 26). As such, this may very well be generally agreed upon but how to distinguish true ethical preferences from apathy or falsity based on thoughtlessness or malice?

There are no foolproof procedures for establishing true preferences no more than a fair procedure of social choice is proof against unfair use. To be sure, if individuals' preferences are selfish enough, 'Making the poor much poorer and passing on half the plunder to the rest would seem to be both wasteful and inequitous, but MMD [method of majority decision] strongly supports such a change' (Sen, 1977b; 1982, p. 163; [...] added). The problem of social choice is essentially an ethical problem. 'All that we can do is to work out the implications of different values, to assess their acceptability by making clearer their consequences and to strike a balance between them as well as we can' (Pattanaik, 1971, p. 171).

7. CONCLUDING REMARK

Traditional SC theory deals with amalgamating individual preferences on a given set of well-defined alternatives into a social preference on those alternatives. In that context 'fairness' is defined as a set of constraints imposed on such an amalgamation. In complex decision problems - concerning, for instance, technological facilities involving grand-scale societal and environmental impacts - using methods of decision making considered in traditional SC theory requires three preliminary steps:
(i) Complex decision problems are not defined as a problem of choosing a best alternative from a given set of well-defined alternatives. Consequently, the decision problem at hand should be structured so that a set of well-defined alternatives emerges. A process of consultations and deliberations is a natural way to obtain such a structure. It is particularly advisable to structure, if possible, the problem in such a way that the emerging set of alternatives is an efficient set. In such a set the alternatives are ordered and, within that order, the attractive and unattractive attributes of the alternatives are ordered the same while the unattractive attributes increase relatively faster than the attractive ones. An efficient set gives easily rise to single-peaked preferences which considerably facilitates step (ii). Furthermore, an efficient set provides clarity on

the nature of the conflict and the trade-off to be made, and makes it accesible for public discussion.

(ii) Amalgamating a set of individual preferences requires a specification of the concept of fairness. In its turn, this specification requires a compromise on the constraints seemingly characteristic for fairness that are to be imposed on the rule of social choice.

(iii) The individual preferences presented for amalgamation should be ethical or social preferences rather than personal preferences solely directed at personal well-being. This requires a sense of community and shared responsibility which may be fostered by the process of deliberations in step (i).

Step (i) requires and deserves the bulk of the effort. Ideally, it results into an alternative that all interest groups accept. Creating such an alternative is exhausting and painful but in decision making it is the supreme good. And if parties are powerful enough, it is the only good.

APPENDIX

Let $Z = \{v,w,x,y,z...\}$ and $N = \{i,j,...\}$ be nonempty sets of, respectively, objects and individuals. Let R be a binary relation on Z and let P and I also be binary relations on Z defined as

xPy iff xRy and not yRx,
xIy iff xRy and yRx.

These definitions imply that $R = P \cup I$.
The relation R may be:

(i)   *complete*: xRy or yRx for all $x,y \in Z$
(ii)   *transitive*: xRy & $yRz \implies xRz$ for all $x,y,z \in Z$
(iii)   *acyclic*: vPw & wPx &...& $yPz \implies$ not zPv for all $v,...,z \in Z$
(iv)   *antisymmetric*: xRy & $yRx \implies x=y$ for all $x,y \implies Z$

We interpret R as *weak preference*, P as *strict preference*, and I as *indifference*. We say that R is a *weak order* iff R satisfies (i) and (ii), while R is a *linear order* iff R satifies (i), (ii) and (iv).

Let S be a nonempty subset of Z. We say that x is a *best element* of S with respect to R iff

$y \in S \implies xRy$ for all $y \in S$.

We let $C(S,R)$ denote the *choice set* of S, that is, the set of all best elements of S with respect to R. Sen (1970, lemma 1.1) showed that for a given R the properties (i) and (iii) are necessary and sufficient for the existence of a choice set $C(S,R)$ for every nonempty subset S of Z. Provided $C(S,R)$ is nonempty for every nonempty subset S of Z we say that $C(S,R)$ constitutes a *choice function* on Z.

If for given Z and N we have an individual weak order of preference R_i on Z for every $i \in N$, the set $\{R_1,...,R_n\}$ is called a *profile of preference*. Let F be a rule that assigns to every such profile a binary relation R on Z so that, say,

$$R = F\ (R_1,...,R_n).$$

If R is a weak order, then F is said to be a *social welfare function* (SWF). If R only satisfies (i) and (iii), then F is said to be a *social decision function* (SDF).

REFERENCES

Arrow, K.J. (1963). *Social choice and individual values*. New York: Wiley. First edition: 1951.

Aschenbrenner, K.M. (1981). Efficient sets, decision heuristics, and single-peaked preferences. *Journal of Mathematical Psychology, 23*, 227-256.

Bell, D.E., Keeney, R.L. and Raiffa, H. (Eds.).(1977). *Conflicting objectives in decisions*. New York: Wiley.

Bezembinder, Th. (1981). Circularity and consistency in paired comparisons. *British Journal of Mathematical and Statistical Psychology, 34*, 16-37.

Blair, D.H., and Pollak, R.A. (1983). Rational collective choice. *Scientific American, 249*, 76-83.

Chen, K. and Mathes, J.C. (this volume). Value oriented social decision analysis: a tool for public decison making on technical projects.

Condorcet, M. (1785). *Essai sur l'application de l'analyse à la probabilité des décisions rendues à la pluralité des voix*. Paris: Imprimerie Royale.

Coombs, C.H. and Avrunin, G.S. (1977a). Single-peaked functions and the theory of preference. *Psychological Review, 84*, 216-230.

Coombs, C.H. and Avrunin, G.S. (1977b). A theorem on single-peaked preferencefunctions in one dimension. *Journal of Mathematical Psychology, 16*, 261-266.

Fishburn, P.C. (1970). The irrationality of transitivity in social choice. *Behavioral Science, 15,* 119-123.

Kelly, J.S. (1978). *Arrow impossibility theorems.* New York: Academic Press.

Krantz, D.H., Luce, R.D., Suppes, P., and Tversky, A. (1971). *Foundations of measurement. Volume I. Additive and polynomial representations.* New York: Academic Press.

Luce, R.D. and Raiffa, H. (1957). *Games and decisions: introduction and critical survey.* New York: Wiley.

Möller, R. (1983). *Interpersonelle Nutzenvergleiche: Wissenschaftliche Möglichkeit und politische Bedeutung.* Göttingen: Vandenhoeck and Ruprecht.

Narens, L. and Luce, R.D. (1983). How we may have been misled into believing in the interpersonal comparability of utility. *Theory and Decision, 15,* 247-260.

Pattanaik, P.K. (1971). *Voting and collective choice.* Cambridge: At The University Press.

Phillips, L.D. (1984). A theory of requisite decision models. *Acta Psychologica 56, 29-48.*

Phillips, L.D. (this volume). Requisite decision modelling for technological projects.

Rawls, J. (1971). *A theory of justice.* Cambridge (Mass.): Harvard University Press.

Richelson, J. (1975). A comparative analysis of social choice functions. *Behavioral Science, 20,* 331-337.

Richelson, J. (1978). A comparative analysis of social choice functions, II. *Behavioral Science, 23,* 38-44.

Sen, A.K. (1969). Quasi-transitivity, rational choice and collective decisions. *Review of Economic Studies, 36,* 381-393. Reprinted in A. Sen (1982).

Sen, A.K. (1970). *Collective choice and social welfare.* San Francisco: Holden-Day.

Sen, A.K. (1977a). On weights and measures: informational constraints in social welfare analysis. *Econometrica, 45,* 1539-1572. Reprinted in A. Sen (1982).

Sen, A.K. (1977b). Social choice theory: a re-examination. *Econometrica, 45,* 53-89. Reprinted in A. Sen (1982).

Sen, A.K. (1979). Personal utilities and public judgements: or what's wrong with welfare economics? *Economic Journal, 89,* 537-558. Reprinted in A. Sen (1982).

Sen, A.K. (1980). Equality of what? In: S. McMurrin (Ed.), *Tanner lectures on human values.* Cambridge: Cambridge University Press.

Reprinted in A. Sen (1982).

Sen, A.K. (1982). *Choice, welfare and measurement*. Oxford: Basil Blackwell.

Sen, A.K. (1985). *Commodities and capabilities*. Amsterdam: North-Holland.

Sen, A.K. (1986). Foundations of social choice theory: an epilogue. In: J. Elster and A. Hylland (Eds.), *Foundations of social choice theory* (pp. 213-248). Cambridge: Cambridge University Press.

Wilke, H.A.M. (this volume). Promoting personal decisions supporting the achievement of risky public goods.

Williamson, O.E. and Sargent, T.J. (1967). Social choice: A probabilistic approach. *Economic Journal, 77*, 797-813.

PROMOTING PERSONAL DECISIONS SUPPORTING THE ACHIEVEMENT OF RISKY PUBLIC GOODS

HENK A.M. WILKE

Department of Psychology
University of Groningen

1. INTRODUCTION: THE NATURE OF SOCIAL DILEMMAS

Many societal problems are social dilemmas. Findings collected in the area of experimental games may be of relevance to people confronted with practical social dilemmas. In this chapter it is explained what social dilemmas are, and some examples of experimental social dilemmas are given. Thereafter, several findings collected in this area will be reported. Subsequently it will be shown that these findings have practical implications. Considerations of power and fairness appear to be important. The final part of the paper focusses explicitly on power and allocation processes. Practical implications for the management of social dilemmas in technological projects are discussed; they are summarized in a flowchart.

1.1 Maintenance of a public good and the prevention of a collective bad

Situations in which private interests are at odds with public interests constitute an important class of societal problems. Undoubtedly, Hardin's 'Tragedy of the Commons' provides a classical example (Hardin, 1968). It describes how individual farmers have access to grazing grounds that are held in common by all. Each can make a personal profit by adding successive cattle to the commons and each continues to do so. The choice of adding an extra animal does involve costs in terms of pasture consumed, but these costs are consumed by the collectivity. A tragedy comes into being because every farmer is inclined to increase his own herd, while leaving the costs involved to the collectivity. But as a result, in the aggregate, far greater costs are likely to be generated than individually absorbed benefits. Eventually the commons will be destroyed. In the terminology of game theory, adding extra animals is the dominating strategy for all individuals. However, the resulting collective outcome is suboptimal,

39

Ch. Vlek and G. Cvetkovich (eds.), Social Decision Methodology for Technological Projects, 39–59.
© *1989 by Kluwer Academic Publishers.*

i.e., nobody wants the commons to be destroyed. Selfish individuals, however, are led to behaviors that produce this deficient outcome.

Hardin's metaphor is an example of a group of individuals who are faced with the problem of how to maintain their collective good and who - if they give in to their individual 'greed' - actually produce a collective bad. What is necessary to prevent this undesired product from a collective point of view, is that individuals restrain themselves. This is not an easy affair. Any individual may consider two possibilities. First, if others do exercise restraint, one can personally enjoy the fruits of their restraint without having to contribute to its costs. That is, by being a *free rider*, one profits from the fact that others prevent the collective bad, or put differently: that others contribute to maintaining the public good. Second, one may consider the possibility of being a *sucker* who makes a cost when nobody else does, perhaps ending up carrying the total cost. But why should one exercise restraint when others might not?

Hardin's metaphor refers to societal problems in which individuals are faced with the problem of how to maintain a public good, a desired collective product, and how to prevent a public bad, an undesired collective product. A more recent example of such a social dilemma is offered by environmental pollution. It is in everybody's interest to maintain an unpolluted environment and to avoid a polluted one. However, the question is who will make the costs and efforts to maintain an unpolluted environment? Nobody likes to be the only one who takes measures to achieve this, and everyone likes all others to make the necessary costs. In other words, because for each individual decision maker the polluting choice is dominant, a suboptimal outcome, a collective bad, may be produced.

Also other examples may be given: living in a quiet environment may be a public good, however it is always more attractive that others reduce *their* noise level and since this is so for all participants (people living in the same house, industrialists sharing an industrial area), an extremely noisy environment may be the undesired consequence, the collective bad. Also on an international level social dilemmas may be indicated. For example, clean rivers may be conceived as a collective good. However, who is making the costs to maintain clean rivers? Since it is more desirable that neighboring nations take the costly measures involved, a collective bad may be the consequence.

1.2 Provision of public goods in order to remove a collective bad

The social dilemmas dealt with before refer to situations in which the initial situation is desirable from a collective point of view. But it asks for costly individual actions to keep it desirable. Below, some social situations are exemplified in which the initial situation is an *un*desirable one. This demands costly individual actions to make the situation desirable from a collective point of view. In the following, two examples of the latter situation will be offered. A community can be faced with the problem whether it will continue to employ a traditional electric power plant, or whether it will install a nuclear power plant. If we assume that for the community at large using traditional forms of energy is a public bad, whereas the use of nuclear energy would be a public good, the problem is who is to make the costs in order to install the nuclear plant. If all want to keep their individual profits from the current situation, the public good will not be provided.

A similar problem may be involved in the case of Liquefied Petroleum Gas (LPG). Having LPG available may be considered a public good when for the society at large the advantages are greater than the disadvantages. It is public, because everyone has access to LPG, i.e., every member of the public is allowed to purchase LPG at a low price. But here, too, the question is who is to incur the costs, or stated differently, who will profit more. Again a social dilemma comes into being because any participant feels more attracted to take the advantages and letting others incur the costs. We all like to have LPG in our cars, but we dislike having the route for LPG tank trucks pass our front doors, because this would expose us to the hazard of a tank explosion.

Nuclear energy and LPG may be considered as desirable from a collective point of view. These collective goods may be called risky public goods because they can only be provided when the individual participants are willing to incur the inevitable risks and costs involved.

1.3 Two perspectives

Until now two types of social dilemmas have been distinguished. They have the following two properties in common: (1) In all circumstances for each individual it is most attractive to make a 'defective' choice, i.e., each person has a dominant strategy available that yields the person the best pay-off in all circumstances; (2) If all individuals make a defective choice, a public bad will be produced, or maintained. In that case a collectivity is worse off than if all had made a 'cooperative' choice, choices which produce a collective good. That is, the collective choice of individually dominating strategies results in a deficient collective outcome, a result that is less preferred by all persons than the result which would have occurred if all had not chosen their dominating strategy.

The two types of situation differ in the perspective taken into account. In Section 1.1 a situation has been described in which initially a public good exists whose continuation is threatened by individuals promoting a collective bad. In Section 1.2 the initial situation implies a collective bad which may be removed when participants make cooperative choices, which eventually produces a public good.

2. SOME EXPERIMENTAL GAMES

Having distinguished two types of social dilemmas, referring to the maintenance of a public good and to the provision of a public good, respectively, it will now be shown how by means of experimental games, these two types of social dilemmas may be simulated and investigated for the personal and social behaviors involved.

2.1 Maintenance of a public good

A so-called *Take Some Game* (see Dawes, 1980; Orbell and Dawes, 1981) simulates a social dilemma situation in which participants have to maintain a collective good, but are tempted to produce a collective bad. In this game one 'takes some' from the collective by making choices analogous to those involved in the decision to pollute. An example is offered by the game employed by Rutte, Wilke and Messick (1987):

> In the Take Some Game subjects learn that there are 27 guilders available for the six-person group. The following instructions are given. Each individual member is allowed to take between 0 and 9 guilders from the total amount. The task of the group, however, is to take altogether at most 27 guilders. Each member has to decide for him- or herself how

many guilders (s)he wants to take. If the total amount
requested exceeds 27 guilders, all group members receive
nothing. If the group's collective take is equal to or less
than 27 guilders, each group member may keep whatever
(s)he has taken before.

2.2 Provision of a public good

A so-called *Give Some Game* simulates a social dilemma situation in
which participants have to provide a public good in order to avoid a
collective bad, i.e., 'give some' increases the likelihood that a
collective good will be provided, whereas the alternative 'keep some'
increases the chance of a collective bad, a consequence analogous to
the decision of not having LPG. The experimental procedures of a Give
Some Game are as follows: before the experiment each of six group
members receives nine guilders from the experimenter. The following
instructions are given:

> 'The task of the group is to give back to the experimenter
> at least 27 guilders. Every group member has to decide for
> him- or herself how many guilders (s)he wants to give up.
> If the group fails to turn in at least 27 guilders, all group
> members are fined by losing their entire amount of 9
> guilders. If the group members together succeed in returning
> at least 27 guilders, every group member may keep whatever
> is left from the original of 9 guilders' (Rutte et al., 1987).

2.3 Comparison between Take Some and Give Some games

It should be noted that the pay-off structure for the Take Some and
Give Some games are numerically identical. The only difference is the
afore-mentioned difference in perspective. In Take Some games
subjects may maintain a public good, which is threatened by those who
take too much. In Give Some games subjects may eventually produce a
public good, a public good which will not be provided if one gives too
little. Rutte et al. (1987), comparing the behaviors of subjects in both
games, observed no difference, suggesting support for Dawes'
supposition that the logic of the collective good and collective bad
works in the same way, i.e., individuals are inclined to burden others
with the costs of the maintenance or the provision of collective
goods.

3. FINDINGS

The foregoing analysis suggests that public goods will never be chosen, because individual interests forbid it. However, we all are aware that such goods are actually provided: LPG, nuclear power plants, armies, and labor unions are a few examples. Consequently the question arises why public goods are supplied in spite of individual restraint to do so. Answers to this question provide suggestions for the management of other public goods dilemmas. In the following some factors will be discussed which facilitate the resolution of social dilemmas. The research findings have been collected in the area of experimental social dilemmas, a variety of game settings having more or less the same outcomes' structure as described before (see also Dawes, 1980; Orbell and Dawes, 1981; Messick and Brewer, 1983; Wilke, Messick and Rutte, 1986).

3.1 Monetary incentives

Kelley and Grzelak (1972) varied the relative attractiveness of a defective choice with regard to the cooperative choice. It appears that when the difference in financial outcomes between a defective and a cooperative choice is small, individuals make relatively more cooperative choices than when this difference is relatively large, indicating that by increasing financial incentives for cooperative behavior, cooperative behavior is strengthened. However, even when the rewards for a defective choice are much larger than the ones for a cooperative choice, a considerable number of group members still decide to make the cooperative choice. Other research findings (see Marwell and Ames, 1979) indicate that in social dilemma situations there always are some subjects who make cooperative choices, and collective goods are more often provided than one would expect when participants were strictly rational - as assumed by game theory. Marwell and Ames (1979) conclude from their work that a stringent free-rider hypothesis, stating that everyone makes a defective choice while leaving it to others to provide the public good, is not supported by empirical evidence.

3.2 Extra-game motives

Given the above-reported results the question arises why some subjects are inclined to help provide the public good in spite of the dominance of a non-providing or defective choice. The following section summarizes some of the effects of the so-called 'extra-games motives', psychological processes which are not directly related to the outcome structure of the games involved.

A variety of factors have been shown to affect cooperative behavior. Since there are various thorough reviews available (see Dawes, 1980; Messick and Brewer, 1983; Wilke, Liebrand and Messick, 1983), it is sufficient to mention only the most relevant ones.

- Communicating groups are more cooperative than non-communicating groups;
- Inducing a sense of 'groupness' favors the provision of a public good;
- Making choices publicly evokes more cooperative behavior compared with a situation in which participants can not be held responsible for their behavior;
- Cooperation is greater in small than in large groups;
- Personality differences play a role: subjects who have a cooperative disposition are more inclined to make a cooperative choice;
- The expectation that other group members will behave cooperatively strengthens one's own tendency to be cooperative.

The extra-game motives just mentioned are psychological side payments which do not belong to the proper outcomes structure of the games. Unfortunately we do not know (yet) how monetary and psychological incentives work in combination (see Kelley and Thibaut, 1978). It is even possible that a combination of monetary and psychological outcomes may lead to a cancellation of the social dilemma character of the situation. This is the case when for each individual the cooperative choice is made more attractive than the defective choice, making the game one of pure coordination for which the conflict between individual and public interests is absent.

For those who want to increase the chance that a public good will be provided the lesson seems to be that participants must be induced to believe that to provide a public good is also beneficial for each of them individually. In reality, information, guidance and propaganda may affect this aim. In sum, social dilemmas are framed by people. Given the formal structure of the game strictly individual rationality can not explain why public goods are provided. However, through framing, i.e., social psychological construction, social dilemmas may be turned into coordination settings in which it is 'psycho-rational' for each individual to contribute to the provision of the public good. For example, the provision of an unpolluted environment such as having clean streets and parks, although unlikely from a strictly game-theoretical point of view, may be encouraged when a society at large is divided in smaller groups, each having its own area of responsibility, when communication among group members is encouraged, when the expectation is strengthened that other group members are already taking care of clean streets and parks and when socialization processes stress the importance of societal instead of personal benefits.

3.3 Group heterogeneity

The foregoing results have mainly been collected with the help of experimental games in which individuals have primarily equal interests, i.e., individuals are homogeneous as to their interests. Olson (1965) argues that group heterogeneity is favorable for the provision of collective goods, since people having greater interests are more likely to produce the collective good than people who have smaller interests. Hardin (1982) gives the example of an industrialist willing to pay $27,000 in campaign contributions to lobby for a tax change worth $15 million to him, even though many others would gain a total of $150 million by that change.

Two clearcut bases for heterogeneity in groups may be distinguished (see also: Oliver, Marwell and Texeira, 1985): interests and resources. First, group members may differ with regard to their desire to provide a collective good. For example, LPG suppliers have a greater interest in providing the risky transport of LPG than owners of cars which run on regular gas or people who live close to LPG gas stations (see Kuyper and Vlek, 1984, for corresponding differences in risk evaluation). And, although we all want clean air, those among us who suffer from emphysema want it more. In line with the foregoing, some of the extra-game motives discussed before may affect differential desires to provide the public good. For example, it may be assumed that people who have a cooperative disposition attach greater value to the provision of the public good than others, who do not have such a disposition.

Secondly, some people may have more resources available to contribute to the public good than others. For example, some of us have more energy or money available to take care of an unpolluted environment than others. The size of these resources may be uncorrelated with the *desire* to contribute to the collective good. Both bases of group heterogeneity may be induced in a social dilemma game by varying independently the resources of the participants (e.g., subjects having previously collected many or few resources) and the size of the stake one has in achieving the collective good.

Empirically, not much is known about the effect of group heterogeneity. Schonfield (1975) discusses how public goods may be provided by resourceful actors who form a coalition, while partly monopolizing the public good outcomes involved, and who collectively take care of the so-called 'critical mass', i.e., the number of cooperators necessary to provide the public good.

Which practical implications can be derived for those who want to advance the availability of risky public goods? Two notions seem to be of importance: heterogeneity and critical mass. As we have seen above, heterogeneity promotes the provision of public goods. In order to estimate whether a public good might be provided, heterogeneity should be assessed beforehand. However, that is not all there can be done. Heterogeneity of interests may also be increased by extra inducements to the actors involved, e.g., by granting them psychological or monetary side payments, as well as by bringing into play other actors, who before were no part of the public choice game, but who by their wealth or desire could turn the balance to the better.

The second important notion is 'critical mass'. Managers of risky public goods and politicians may consider beforehand the critical mass involved. For example, in parliament the critical mass consists of the majority of votes, i.e., a majority of votes is sufficient to make a decision. For other public good decisions, it is often rather difficult to assess beforehand the actors of relevance and their influence on the final decision, while also the decisional procedures which determine whether the mass is critical or sufficient to provide a public good are not well defined. However, one may realize that ambiguities leave space for an active approach. By lobbying, by offering information as well as side payments and by influencing procedural rules, public goods may be promoted in an active way. One of the ways to influence these procedures is to establish a superimposed authority, a solution of social dilemmas to be discussed in the following.

3.4 Superimposed authority

The characteristics of social dilemma situations have long been known.
In the 17th century Hobbes argued that a social dilemma can only be
solved by the instalment of a superimposed authority. This solution
implies that the freedom of individual decision makers is restricted and
that one powerholder takes the decisions on behalf of all people
involved. Spinoza, sharing basically the same insights, pointed out that
the power to decide can also be handed over to a democratically
chosen body of 'super' decision makers. In the same vein, Hardin
(1968) expressed the belief that social dilemmas can only be solved by
some kind of mutual restraint and that solutions which do appeal to
individual cooperation are doomed to be unsuccessful. Quite recently
some research findings have become available about the conditions
under which one prefers to have a superimposed authority or a leader
and how one tends to respond to a once elected leader.
   - In general one has a strong reluctance to hand over one's
     decisional freedom to a superimposed authority (Rutte and Wilke,
     1985). Apparently handing over one's decisional freedom is
     perceived as so costly that considerable compensations must be
     made available.
   - When confronted with the question whether one would prefer a
     leader above maintaining one's own freedom of choice, the
     history of the group appears to play a role (Messick et al., 1983;
     Rutte and Wilke, 1984; Samuelson, Messick, Rutte and Wilke,
     1984). Two conditions seem to be of importance. First, when
     previously the group without a leader is unable to maintain a
     public good, one is subsequently more inclined to accept a leader
     than when the group on its own succeeds in maintaining a public
     good. Second, when previously in the group without a leader the
     outcomes among group members are distributed in an unfair way,
     a stronger preference to elect a leader is observed than when the
     outcomes are distributed equally, i.e., fair among group members.
   - An other research question is how one responds to an elected
     leader. Two types of responses by subordinates have been
     investigated, viz. social support for a leader, i.e., the tendency to
     re-elect a once-appointed leader, and compliance to directions of
     the leader, who advises about how much each participant should
     harvest in a sequential Take Some game. As for the endorsement
     of a leader it appears that more support for a leader is expressed
     when (s)he succeeds in achieving the collective good than when
     (s)he fails to do so (Wilke, Liebrand and De Boer, 1986).

Moreover, when the leader allocated outcomes in a fair way (s)he received more support then when (s)he made unequal allowances to her/his subordinates (Wit and Wilke, 1984). As for compliance to the advice of a leader, it is observed that the perceived competence of the leader plays a crucial role. One complies more easily to given advice when it is known that the advisor is successful in directing his or her subordinates and when one expects him to be superior in this respect beforehand.

Summarizing the foregoing in a more practical sense, we may conclude that the instalment of a superimposed authority does not seem to be an easy affair, since people dislike to hand over individual decisional freedom to a decision making body or to an individual powerholder. This reluctance can be overcome, however, when one is sufficiently compensated for such a loss of free access to a public good. One accepts a certain loss of control when faced with a situation in which one is threatened with a collective catastrophe or when there are large social discrepancies in the allocation of outcomes. For those who want to establish a superimposed authority these results suggest that the framing of a situation as one in which large unequalities or an impending catastrophe is at stake, might be successful in achieving their aim.

In the same vein, a once-established authority should prove that it can operate successfully and that it is able to avoid the rise of strong feelings of unfairness. When this is not the case the specific authority will be replaced by an other one. One can imagine that if subsequent authorities are not able to manage the public good succesfully or if they are not able to keep (feelings of) unfairness below a tolerable level, the trust in any authority might be shaken. At that moment participants will be inclined to turn away from having any authority at all, and the social dilemma character of the situation is re-established in a more pronounced way. Every individual is again inclined to pursue individual interests, and a collective catastrophe may be expected even more strongly than before, when the hope for a collective good was still relatively high. The lesson seems to be that authorities do well to realize the two afore-mentioned goals. If not, having authorities which fail in sequence diminishes the long run likelihood that public goods will be provided at all.

4. POWER

After the overview of factors facilitating the provision of public goods, it seems worthwhile to pose the question of what the common denominator of these solutions is.

In my opinion the answer is 'power'. Power plays a role in situations implying a certain conflict. This is clearly the case for social dilemmas, since they imply a conflict between (some) individual and group interests. When an actor or a group is able to supply the public good against the desire of others who are not willing to do so, the actor or the group is assumed to have exercised power. More generally, 'A has power over B, to the extent that A can get B to do something B would not otherwise do' (Dahl, 1957).

Several bases of power, i.e., means by which a powerholder may affect the behavior of recipients, may be distinguished. French and Raven's (1968) well-known classification distinguishes the following five bases.
- *reward* and *coercive power*: the ability to reward or to punish;
- *legitimate power*: the recipient acknowledges that the powerholder has the right to influence him or her and (s)he has the obligation to comply;
- *referent power* is present when a power recipient identifies with a powerholder and tries to behave like him or her; in this case a powerholder may be even unaware that (s)he is in fact a powerholder.
- *expert power* is based on special knowledge attributed by the recipient to the powerholder.

It is not difficult to demonstrate that power plays a crucial role in the foregoing analysis of solution strategies for social dilemmas. First, by power the provision of public goods may be established. Providing extra rewards to someone who is initially risk avoidant, e.g., as in the case of LPG, might turn him or her into someone who is risk acceptant. When an isolated group of individuals fails to help provide the public good, a superimposed authority of their own might be accepted by his or her fellow group members. Moreover, extra-game motives may be aroused by a combination of expert and referent power. The presented ideas about heterogeneity of interests suggest that affluent group members may help to provide the public good. Second, by means of power the provision of public goods may be maintained. A superimposed authority is supported when (s)he contributes to the group success, i.e, as long as (s)he promotes the collective good. Moreover, one complies to an advisor when the latter has manifested superior competence, i.e., expert power, and when superior rewards, i.e., reward power, become available.

The above reasoning suggests that risky public goods are provided and maintained by power. However, our analysis also suggests that the concept of power refers to several power bases and the problem is that for concrete situations it is rather difficult to specify beforehand which power base will be effective. This conclusion also seems to hold for field studies. In a recent study Hisschemöller, Midden and Stallen (1985, see also Hisschemöller and Midden, this volume) describe four cases for which the provision and maintenance of risky public goods are analyzed. They make a distinction between several 'power bases': coercion, compensation, information or legitimation. They indicate that in different cases different combinations of 'power bases' are involved. Their interpretation of these case studies leads them to conclude that participation of all parties involved increases the likelihood that decisions will be perceived as legitimate and fair.

Such types of evaluation study may be very useful in determining which power base may be effective in which specific situation. Moreover, experimental as well as field data seem to suggest that until now this type of knowledge is rather scarce. It would, however, seem to be most useful for practical decision makers. In the same vein, it may be concluded that general normative models of power (e.g., Mulder, 1972; Cook and Emerson, 1978), although useful in themselves, need to be tuned towards specific situations. As will be argued in the following section, this also holds for the concept of fairness.

5. FAIRNESS

Earlier we have seen that in experimental games the concept of fairness, equity or justice plays a crucial role in the establishment and enactment of a superimposed authority. It appears that one prefers to have an authority when in a leaderless group the outcomes are unequally distributed among the group members. Social support for a leader is stronger when the latter allocates outcomes in a fair way. Moreover, it was observed that group members when made leader, strived to allocate outcomes equally among all group members. These data are important because they make clear that in experimental settings in which subjects have equal resources, they prefer outcomes to be allocated equally. One may doubt, however, whether this conclusion is transferable to other experimental situations and to real-life situations. There are several reasons - partly theoretical and partly empirical - for this doubt.

To understand this, let us briefly consider how the idea of fairness developed within social psychology. The first conceptualizations were more or less literal translations from economic theory (Adams, 1965) where it is assumed that market actors come to an agreement when they balance their investments and outcomes. Indeed, it was found quite often that in performance situations individual outcomes matched (were proportional) to one's contributions (effort, ability and so on; see Wilke, 1983). Later, other fairness rules were discovered (see Törnblom and Johnson, 1985). Individual need also appears to be an important cue for the 'fair' allocation of outcomes. In situations in which procedures are made explicit, procedural equality or equal opportunity of treatment appears to be pursued.

Being confronted with these rules, researchers focussed their attention on the question of which rule operates when. For example, Brickman, Folger, Goode and Schul (1981) demonstrated that members of a group will consider equality of outcomes more just than equality of opportunity when information about individual contributions to the group performance is lacking.

Messick and Sentis (1983) have succinctly summarized the state of affairs arguing that equality seems to be at the heart of the concept of fairness and that the question is not whether equality is important, but on what dimension equality should be established. The problem of finding fair arrangements is not one of equality versus inequality but one of equality with regard to what. Thus, the dimension of comparison seems to be a key element which is determined in a situation-specific way. Nowadays social psychologists in this area are performing experiments to trace the contingencies of fairness rules and types of situation. For several reasons this does not appear to be an easy affair.

First, the dimensions on which equality might be established are subjective, psychological dimensions. How can people tell they are equal in their need or in their invested effort? A similar problem arises concerning the assessment of outcomes: how much satisfaction is equal to how much financial compensation? Second, how am I to infer the need or risk involved for somebody else? A fundamental asymmetry is involved in one's ability to assess one's own position in comparison to that of other people. This is so because one has direct access to one's own subjective states, but those of others can only be inferred from their behavior (see also Bezembinder, this volume). A third problem is that people quite often make judgments which are assumed to be beneficial to them in the long run. For example, a person may exaggerate the riskiness of his or her own situation in the hope to receive a larger compensation in the future. A final problem is that people do differ in the dimensions of comparison on which they like to pursue fairness. Research done by Messick and Sentis (1979) indicates that one selects the rule that best conforms to one's preference.

How does the foregoing relate to fairness considerations in collective risky settings? A short walk through the literature on decision making under risk suggests both normative as well as descriptive approaches. Some authors pursue the construction of normative models. For example, Keeney (1978) defined public risk in terms of fatalities to members of the public. He shows that a particular definition of fairness, viz. equality in the degree of risk to each individual, leads to collective risk proneness and subsequently towards a catastrophe. In an other analysis Keeney (1983) develops models for three dimensions or rules of fairness, viz. minimizing the expected number of fatalities, promoting an equitable distribution of goods, and avoidance of a catastrophe. By assigning values with the help of adjusted utility functions he elegantly appraises the consequences of these rules or dimensions for making risky public decisions.

Others try to develop descriptive models and sketch the rules or dimensions of fairness which appear to be applied in specific situations. Derr, Goble, Kasperson and Kates (1981) demonstrate the existence of a double standard indicating that public protection is ordinarily set below the level of medically defined hazard, whereas workers' protection is set above this level, exposing workers to known dangers. They also point out that several principles of fairness may be distinguished, principles which have a strong resemblance to the ones mentioned in connection with the social psychological work done in this area. One of those dimensions is compensation, meaning that those workers who run a greater risk should receive larger compensations. From an analysis by Graham and Shakow (1981, see also Graham,-Shakow and Cyr, 1981) on hazard pay for workers, it appears that for workers the connection between risk and payment is not perfect:

privileged (e.g., unionized) workers are compensated to a greater
extent than non-unionized workers, who earn relatively lower wages.

A broader approach is suggested in three recent studies. From Kuyper
and Vlek (1984) it appeared that the kind of interest someone has in a
particular set of activities affects the way of conceptualizing these
activities in terms of relevance and favorableness. For example, LPG
company managers, presumably having a more direct interest in the
use of LPG, rated statements like driving a car running on LPG,
storage of LPG at a gas station close to a built-up area and
transporting LPG in tank trucks, more favorable than did civil
servants charged with environmental protection. Hisschemöller et al.
(1985, see also this volume) describe four cases in which public hazard
was involved. Their systematic analysis gives rise to the conclusion
that a combination of fairness of compensation and equal opportunity
of the people involved, seems to be most suitable.

Rayner and Cantor (1987) argue that the rules of fairness one employs
refer to the type of social organization to which one belongs. They
investigated the acceptability of new nuclear power technologies and
interviewed three constituencies: utility companies, state public utility
commissons and public interests groups critical of nuclear power. Their
results suggest that due to their organizational background these
constituencies differ widely. For example, members of public utility
commissions are more concerned about the problem of whether nuclear
power plants are economically feasible and whether the utility has
selected the correct technological options, while interest groups are
more concerned about the consent of affected parties and about the
distribution of liabilities. Raynor and Cantor contend that the ways in
which these constituencies differ suggest implicit agendas of interest
that make it difficult for them to understand one an other. This
approach may appear to be promising because it may uncover both the
specific rules involved as well as their organizational anchoring
(Kuyper and Vlek, 1984).

To summarize, in the area of technological risk - e.g., nuclear power
or LPG risk - we find that there are elegant normative models as well
as tentative descriptive models about the way in which risk is
experienced and how risky decisions should be, or are, implemented in
concrete hazardous circumstances. The wrong way to deal with the
observed gap seems to be to propose to neglect the development of
normative models. These models are not only elegant but extremely

useful to sketch the consequences of certain allocation properties. It might be argued (see Keeney, 1983, 1984) that when in specific circumstances the descriptive properties which operate in a specific situation are clearly specified, in order to assess the consequences these operating properties may be built into a relevant normative model. On the other hand it should be stressed that before it is possible to advise responsible public agents with regard to the provision and maintenance of risky public goods, more empirical evidence about power and allocation processes in specific situations is necessary.

6. MANAGEMENT OF RISKY PUBLIC GOODS

Decision problems with regard to technological projects very often imply a social dilemma. Concerning the management of these social dilemmas the following four conclusions may be drawn:

(1) Solutions limiting one's free access to the public good are less favored than solutions which do not restrict one's decisional freedom.

(2) Management techniques should be successful and fair.

(3) Whether or not the outcomes of a management technique are considered successful and fair is a matter of subjective assessment by the people involved.

(4) As for the perception of success and fairness, program research is necessary in order to anticipate and evaluate the consequences of management techniques.

These conclusions have been summarized in Figure 1. Since having a superimposed authority (6) is less favored than the inducement of monetary and symbolic incentives (1), it seems logical to start with the latter inducements, referred to as 'SOFT' management techniques. In order to estimate if the people involved perceive the consequences as fair and successful, empirical program research is necessary. If a sense of fairness and success does result (3) the social dilemma is solved (EXIT). If not, one may reconsider the monetary and symbolic inducements (4) in the hope (3) may be achieved.

If it repeatedly appears that monetary and symbolic incentives are of no avail (5), then the introduction of a superimposed authority is advisable (6), a so called 'HARD' management technique. Again the question arises whether the superimposed authority operates successfully and fairly (7). To estimate whether or not this is the case

Figure 1: The management of social dilemmas.

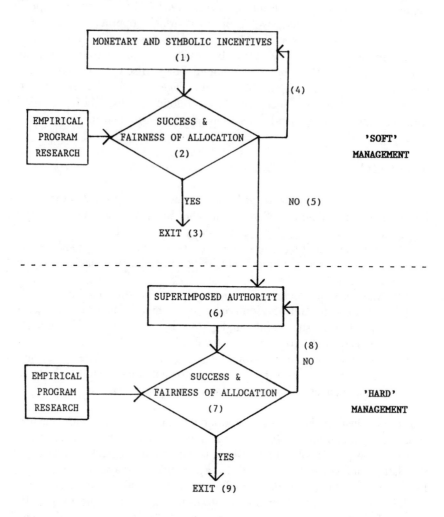

empirical program research is necessary. Again the opinions of the people involved should be evaluated for - as I have argued before - fairness and success are in the eye of the beholder. If it appears that the superimposed authority is perceived as not promoting success and fairness (8), I suggest to reconsider the behavior of the superimposed authority as often as is necessary to establish success and fairness in the eyes of one's constituency.

REFERENCES

Adams, J.S. (1965). Inequity in social exchange. In: L. Berkowitz (Ed.), *Advances in experimental social psychology*, Volume 2 (pp. 267-300). New York: Wiley.

Bezembinder, Th. (this volume). Social choice theory and practice.

Brickman, P., Folger, R., Goode, E. and Schul, Y. (1981). Microjustice and macrojustice. In: M.J. Lerner and S.C. Lerner (Eds.), *The justice motive in social behavior* (pp. 173-202). New York: Plenum Press.

Cook, K.S. and Emerson, R.M. (1978). Power, equity and commitment in exchange networks. *American Sociological Review, 43*, 721-739.

Dahl, R. (1957). The concept of power. *Behavioral Science, 2*, 202-203.

Dawes, R.M. (1980). Social Dilemmas. *Annual Review of Psychology, 31*, 169-193.

Derr, P., Goble, R., Kasperson, R.E. and Kates, R.W. (1981). Worker/public protection: the double standard. *Environment, 23(7)*, 6-36.

French, J.R.P. and Raven, B. (1968). The bases of social power. In: D. Cartwright and A. Zander (Eds.), *Group dynamics* (pp. 259-269). New York: Harper & Row.

Graham, J. and Shakow, D. (1981). Risk and reward: hazard pay for workers. *Environment, 23(8)*, 14-45.

Graham, J. Shakow, D.M. and Cyr, C. (1981). Risk compensation - in theory and practice. *Environment, 25(1)*, 14-40.

Hardin, G.R. (1968). The tragedy of the commons. *Science, 162*, 1243-1248.

Hardin, G.R. (1982), *Collective action*. Baltimore: John Hopkins University Press for Resources for the Future.

Hisschemöller, M. and Midden, C.J.H. (this volume). Technological risk, policy theories and public perception in connection with the siting of hazardous facilities.

Hisschemöller, M., Midden, C.J.H. and Stallen, P.J. (1985). *Het kiezen van locaties voor gevaarlijk (radioactief) afval*. Delft: SIBAS.

Keeney, R.L. (1978). *Equity and public risk*. San Francisco: Woodward-Clyde Consultants.

Keeney, R.L. (1983). Evaluation of mortality risks for institutional decisions. In: P.C. Humphreys, O. Svenson and A. Vári (Eds.), *Analysing and aiding decision processes* (pp. 23-38). Amsterdam, North-Holland and Budapest: Academiai Kiado.

Keeney, R.L. (1984). Ethics, decision analysis and public risk. *Risk Analysis, 4(2)*, 117-129.

Kelley, H.H. and Grzelak, J. (1972). Conflict between individual and common interest in a n-person relationship. *Journal of Personality and Social Psychology*, *21*, 190-197.

Kelley, H.H. and Thibaut, J.W. (1978). *Interpersonal relations*. New York: Wiley.

Kuyper, H. and Vlek, Ch. (1984). Contrasting risk judgments among interest groups. *Acta Psychologica*, *56*, 205-218.

Marwell, G. and Ames, R.E. (1979). Experiments on the provision of public goods. 1. Resources, interest, group size, and the free rider problem. *American Journal of Sociology*, *84*, 1335-1360.

Messick, D.M. and Brewer, M.B. (1983). Solving social dilemmas: a review. *Journal of Personality and Social Psychology*, *4*, 11-44.

Messick, D.M. and Sentis, K.P. (1979). Fairness and preference. *Journal of Experimental Social Psychology*, *15*, 418-434.

Messick, D.M. and Sentis, K. (1983). Fairness, preference and fairness biases. In: D.M. Messick and K.S. Cook (Eds.), *Equity theory* (pp. 61-94). New York: Praeger.

Messick, D.M., Wilke, H.A.M., Brewer, M.B., Kramer, R.M., Zemke, P.E. and Lui, L. (1983). Individual adaptations and structural change as solutions to social dilemmas. *Journal of Personality and Social Psychology*, *44*, 294-309.

Mulder, M. (1972). *Spel om macht*. Meppel: Boom.

Oliver, P., Marwell, G. and Texeira, R. (1985). A theory of the critical mass I. Interdependence, group heterogeneity, and the production of collective goods. *American Journal of Sociology*, *91*, 522-556.

Olson, M. (1965). *The logic of collective action*. Cambridge, Mass.: Harvard University Press.

Orbell, J. and Dawes, R. (1981). Social dilemmas. In: G.M. Stephenson and J.M. Davis (Eds.), *Progress in applied social psychology, Volume 1* (pp. 37-65). Chichester: Wiley.

Rayner, S. and Cantor, R. (1987). How fair is safe enough? The cultural approach to societal technology choice. *Risk Analysis*, *7*, 3-13.

Rutte, C.G. and Wilke, H.A.M. (1984). Social dilemmas and leadership. *European Journal of Social Psychology*, *14*, 105-121.

Rutte, C.G. and Wilke, H.A.M. (1985). Preference for decision structures in a social dilemma situation. *European Journal of Social Psychology*, *15*, 367-370.

Rutte, C.G., Wilke, H.A.M. and Messick, D.M. (1987). The effects of framing social dilemmas as give some or take some games. *British Journal of Social Psychology*, *26*, 103-108.

Samuelson, C.D., Messick, D.M., Rutte, C.G. and Wilke, H.A.M. (1984).

Individual and structural solutions to resource dilemmas in two cultures. *Journal of Personality and Social Psychology, 47,* 94-104.

Schonfield, N. (1975). A game theoretic analysis of Olson's game of collective action. *Journal of Conflict Resolution, 19(3),* 441-461.

Törnblom, K. and Johnson, D.R. (1985). Subrules of the equality and contribution principles: their perceived fairness in distribution and retribution. *Social Psychology Quaterly, 48(3),* 249-261.

Wilke, H. (1983). Equity: information and effect dependency. In: D.M. Messick and K.S. Cook, (Eds.), *Equity theory* (pp. 47-60). New York: Praeger.

Wilke, H.A.M., Liebrand, W.B.G. and De Boer, K. (1986). Standards of justice and quality of power in a social dilemma situation. *British Journal of Social Psychology, 16,* 51-53.

Wilke, H.A.M., Liebrand, W.B.G. and Messick, D.M. (1983). Sociale dilemma's, een overzicht. *Nederlands Tijdschrift voor de Psychologie, 38(8),* 463-480.

Wilke, H., Messick, D.M. and Rutte, C.G. (Eds.). (1986). *Experimental social dilemmas.* Frankfurt am Main: Peter Lang.

Wit, A.P. and Wilke, H.A.M. (1984). Social support for a leader in a social dilemma situation. *Proceedings of the International Association for Research in Economic Psychology: Linking Economics and Psychology. Part 1* (pp. 238-269).

COGNITIVE DIMENSIONS OF CONFLICTS OVER NEW TECHNOLOGY[1]

BERNDT BREHMER

Department of Psychology
University of Uppsala

1. INTRODUCTION

The introduction of new technologies invariably leads to conflict. The nature of these conflicts, however, seems to change as the scale of the new technology becomes larger. The destruction of the spinning machines by the Luddites is easily understood in terms of personal interest. Conflicts over nuclear power and national data banks, on the other hand, are harder to conceptualize in the traditional terms of political ideology or personal interest.

Thus, we find members of all political parties on every side of every large scale technological issue, such as the nuclear power controversy (Vedung, 1979). As a consequence, these conflicts cause considerable problems for the political system and they are not easily resolved in the ordinary political process. Extraordinary means are therefore needed to cope with these conflicts. For example, in Sweden conflict over the nuclear power issue required a special referendum, a method rarely employed.

It is true, of course, that in its later stages, a conflict over new large scale technologies can be modelled as a zero-sum game: one side is interested in implementing the new technology, while the other side is interested in abolishing it, and the gain of the one side is the loss of the other. This does not mean, however, that the conflicts originated because of concern for differential gain. Indeed, this is clearly not the case. For example, those citizens who favor nuclear energy often stand to gain or loose as much as those who oppose it.

Instead, it seems more reasonable to think of these conflicts as *cognitively* based - that is, to assume that they stem from differences in how people think about the new technology. There is nothing surprising about this. A new and untried technology raises a number of

61

Ch. Vlek and G. Cvetkovich (eds.), Social Decision Methodology for Technological Projects, 61–77.
© *1989 by Kluwer Academic Publishers.*

questions: Do we really need it? Will it work? How much will it cost? Is it safe? These are very real cognitive issues, and they require us to forecast such things as our future energy needs and costs, the likely outcome of present research into alternative energy sources, and the likelihood of accidents and their severity.

In most cases, we find that it is hard to obtain a satisfactory answer to these questions. They can not be answered on the basis of experience with the technology in question, nor can the answers be calculated on the basis of scientific theory or fact. Instead, we have to resort to judgment, either our own, or that of experts.

In situations of this kind, we often find that people come up with different judgments. Even worse, we find that the experts differ, and each side seems to be able to find an expert who supports their point of view. Thus, in the nuclear power debate, we find experts who forecast that we will need great amounts of energy that can only be supplied by nuclear plants, as well as experts who think we will not need so much energy.

This leaves the decision maker in a situation where (s)he can not fall back upon unanimous expert opinion, but will have to rely upon his or her own nonexpert judgment. This is an extremely uncomfortable situation, for decisions based upon judgment are hard to justify and subject to different and often unflattering interpretations.

Part of the problem here seems to be that we have some fundamentally incorrect intuitions about the cognitive processes involved in judgment, and an exaggerated faith in our ability to make correct judgments. We believe that when new "facts" are presented, it will be easy to change one's views, and adopt the correct point of view. Recent research on human judgment challenges this view, however. This research has important implications for understanding conflict in general, and conflicts caused by the introduction of new technology in particular.

2. RESEARCH ON HUMAN JUDGMENT

Most of the empirical and theoretical analyses of cognitive conflict have been undertaken within the general framework of Social Judgment Theory (SJT), a general approach to the study of human judgment (Hammond, Stewart, Brehmer and Steinmann, 1975). This is not the place for a full discussion of the SJT approach to human judgment (see Brehmer, 1984, 1987 for recent overviews), but a few words about the general approach are necessary as a background to the discussion of conflict.

Social Judgment Theory is concerned with human judgment in situations which require people to make inferences from uncertain information under conditions where they can not, or do not, derive their answer using some analytical scheme. Social judgment theorists find that under these conditions judgment tends to be a quasi-rational process, composed of both analytical and intuitive elements (see, e.g., Hammond, 1982, Hammond and Brehmer, 1973, and Hammond, Hamm, Grassia and Pearson, 1984, for discussions).

The principal alternative to SJT is Decision Theory (DT), described by Phillips and by Chen and Mathes elsewhere in this volume, in two important ways. The first of this concerns the question of how judgments should be analyzed, the second the problem of uncertainty.

Both SJT and DT are concerned with complex judgments, made on the basis of a number of factors. However, whereas SJT favors a posteriori decomposition of holistic judgments, DT favors a priori decomposition, where the complex judgment problem is decomposed into a number of subtasks, and separate judgments obtained for each subtask, usually for each important factor in the problem. Proponents of SJT claim that holistic judgments are more natural and easier to do than decomposed judgments, and that they give a more realistic picture of a person's actual cognitive system because it forces the person to make the inevitable trade-offs between different factors in context. Proponents of DT, on the other hand, suggest that holistic judgments are too complex, and that people are therefore prevented from giving a true picture of their cognitive systems and preferences if they are forced to make holistic judgments. Both of these positions are reasonable enough, and it seems unlikely that only one of them is correct. Instead, we may assume that whether a priori or a posteriori decomposition should be used depends on the task, e.g., whether the persons being analyzed usually make holistic judgments for the task at hand. It is an important, but sadly neglected task to investigate whether a priori or a posteriori decomposition is the best way of analyzing judgment.

As for uncertainty, SJT is concerned with judgments made under uncertainty, while DT is typically concerned with judgments about uncertainty (Brehmer, 1976). That is, proponents of SJT usually ask their subjects to make point estimates about what they think will happen, while proponents of DT usually give (or ask for) a number of alternative outcomes and they require their subjects to estimate the likelihood of each outcome. Again, SJT and DT are not in opposition, but complementary approaches, and it is an important, but neglected task to decide when the one or the other approach is more useful.

SJT uses linear statistical models to study judgment. Specifically, a linear model is fitted to the relations between a set of judgments and the information upon which these judgments are based, and this model is then used to infer the characteristics of the judgment process. The resulting equation is usually referred to as the person's *judgment policy*, and the procedure used to find the policy is called *policy capturing*. Results from numerous studies (see Slovic and Lichtenstein, 1971, and Libby and Lewis, 1982, for reviews) involving a wide variety of judgment tasks and professional groups (and some nonprofessional groups as well) suggest that

    (1) *judgmental processes are relatively simple*: judgments are usually based upon very little information (3-5 cues), and information from these cues are generally integrated in a linear additive way,

    (2) *the judgment process is inconsistent*, that is, the judgments for a set of cases can not be fully accounted for by a single model, and judgments for the same case may vary from occasion to occasion,

    (3) *there are wide interindividual differences*, both with respect to which pieces of information people attend to, and with respect to the weights they give to the pieces of information they use, and this is true of experts who have years of experience with the task used, as well as of novices, and

    (4) *people have little insight into their judgmental processes*, i.e., after having made a series of judgments, their verbal descriptions of how they made these judgments do not predict their actual judgments.

Thus, the view that judgment enables us to make reliable holistic assessments of large amounts of data finds no support in the results from empirical studies of judgment. However, because we have so little insight into how we make our judgments when we operate in the quasi-rational mode, it has been possible to have a very exaggerated view of the quality of our judgment.

These results from studies of human judgment have important implications for understanding cognitive conflicts. They indicate that these kinds of conflicts will involve cognitive systems that are characterized by inconsistency and lack of insight. We would therefore expect that the parties to such conflicts would find it hard to understand one an other, and that there would be considerable opportunities for misinterpreting the nature of the conflict. Thus, we should not be surprised to find that cognitive conflicts will be hard to resolve. As we shall see, this is indeed the case.

3. RESEARCH ON COGNITIVE CONFLICT

3.1 Laboratory studies of cognitive conflict

Laboratory experiments on cognitive conflict are conducted in two stages. The first stage serves to insure that there are relevant cognitive differences between the persons later to be studied in the second stage, the conflict stage. The experimenter may induce these differences by training, or (s)he may select people with different policies. In the second stage, two or more persons with different judgment policies are brought together and asked to work jointly on a series of problems which require them to exercise their judgment.

The results obtained with this laboratory paradigm (see Brehmer, 1976, for a review of these results) suggest that cognitive conflicts are hard to resolve. Specifically, the results show that people are often unable to reduce their disagreement, even if they are able to eliminate the systematic differences between their policies, because of the inconsistency that develops as they change these policies to reach agreement, and that the level of inconsistency varies with the characteristics of the task, such as the level of uncertainty and the level of complexity. Inconsistency has real behavioral consequences: it prevents interpersonal understanding (Brehmer, 1974; Hammond and Brehmer, 1973), and it is thus not an artifact of the methods used for analyzing the subjects' policies.

These results provide a clear demonstration of the importance of cognitive factors in conflict. They also demonstrate that cognitive conflicts can not be analyzed only in terms of the relations between the persons involved; the nature of the task facing the parties to the conflict will have to be considered as well. These points are usually overlooked in traditional analyses of conflict.

3.2 Applied work

The applied work consists of a series of case studies, where a conflict has been identified, and an attempt has been made to analyze and resolve it. While it is not possible to derive precise predictions about these conflicts since we do not have a detailed description of how they arose and developed, it is nevertheless possible to make general predictions about what we should find in these cases from the laboratory results.

In most of the applied work, there has been some protracted period of conflict before the actual study was done. Thus, we can assume that there has been a period preceding the study which resembles the situation analyzed in the laboratory paradigm. The laboratory results lead us to expect that under these circumstances we should find considerable inconsistency, i.e., disagreement should only partly be due to systematic differences in policy. We should also expect misunderstanding of the nature of the conflict as well as of the policies of the opponents. In addition, we should expect that the level of conflict should be due to the nature of the task.

The last hypothesis can not be tested because in the applied studies undertaken so far, there have been no detailed descriptions of the nature of the task. There is, however, confirmation of the other hypotheses. The Balke, Hammond, and Meyer (1973) study serves as a good example.

In this study six negotiators (three from labor and three from management) who had recently been involved in a long strike at a major chemical company agreed to re-enact their negotiation. They agreed to start the process at the point one week prior to the settlement. At that point, they agreed, four issues remained: the duration of the contract, the increase in wages, the number and use of certain "special workers", and the number of strikers to be recalled.

Each negotiator was asked to evaluate twenty-five potential contracts which differed with respect to the above four factors, and a judgment analysis was performed on each negotiator's judgments. The results showed: (1) that there was disagreement, (2) that inconsistency was an important contributor to this disagreement, (3) that the negotiators had little insight into their own policies, (4) that the understanding of the other side was poor; the negotiators from the one side could not predict the responses of the other side to the contracts, despite the fact that the negotiators on both sides were confident that they understood the position of the other side very well. It is interesting to note that there were considerable differences within the negotiating teams; the three members of the labor team were highly similar in their policies, but the three negotiators for management were quite dissimilar. This was not apparent to either side before judgment analysis.

These results confirm the laboratory findings, and show that cognitive factors may be an important feature also in a situation which is basically a conflict of interest. Clearly, these cognitive factors must have contributed to the problems in the negotiations, and they must have been part of the problem in finding an acceptable contract. Inconsistency and lack of insight will have led the negotiators to send incorrect messages, talking about what was important in one way, and then judging contracts in an other way.

There is no other applied study in which both interpersonal understanding and the contribution of inconsistency to conflict has been assessed. However, all of the studies in which policies have been assessed support the notion that lack of consistency is an important contributor to conflict (Flack and Summers, 1971; Hammond, Rohrbaugh, Mumpower and Adelman, 1977). Thus, there is nothing yet in the applied work to contradict the laboratory results. It seems appropriate to conclude provisionally that the experimental results have some generality also outside the laboratory.

4. THE RESOLUTION OF COGNITIVE CONFLICT

There are at least two possible reasons why cognitive conflicts are not resolved. The first is that people are unable to communicate adequately because of inconsistency and lack of insight into their policies. Finding a way of improving communication might therefore help to resolve conflict. The second possibility is that the persons in conflict have problems coming up with, and evaluating, possible compromise policies because they can not generate and evaluate the implications of possible new policies. We (Hammond and Brehmer, 1973,

see also Stewart and Carter, 1973, and Cook and Hammond, 1982) have tried to solve both of these problems by means of a computerized decision aid called POLICY.

4.1 Communication problems: a computerized communication aid

POLICY relies upon interactive computer graphics. It presents information about a number of cases, accepts judgments from the persons in conflict, analyzes these judgments, and displays the results graphically.

The system shows which information factors were actually used by the persons in conflict, the relative weights given to these factors, the functions relating judgments to cues, and the consistency of their judgmental policies. The system also shows the extent to which conflict is caused by systematic differences and inconsistency.

POLICY, then, not only makes it possible to obtain a precise picture of what the differences between two cognitive systems are; it also makes it possible to communicate what has never been communicated before: the characteristics of entire cognitive systems.

The Balke et al. (1973) study referred to above provides an example of how POLICY can be used to reduce conflict. In this study, four of the negotiators were shown the results of the judgment analysis. This led them to change their policies and to reduce their disagreement. When the negotiators had been able to understand each others' policies from the information supplied by POLICY, they realized that a number of the proposed contracts were acceptable to both sides, something that they had not been able to do on the basis of the unaided discussion. This suggests that the dispute could have been shortened considerably if POLICY had been available during the actual negotiations. It also emphasizes that at least part of the problems in cognitive conflict has to do with people's inability to communicate, as was suggested by our initial analysis.

4.2 Finding acceptable solutions

Better understanding and communication may not always be enough to resolve conflict; the problem of finding an acceptable solution still remains. Finding such a solution involves trying out new policies and exploring their consequences. This is not an easy task. A series of

studies (see, e.g., Brehmer, Hagafors and Johansson, 1980) show that people have considerable problems in using a given policy to derive judgments, especially when the policy is complex. Consequently, people are likely to make mistakes when trying to understand what a proposed policy means for concrete cases, and they may therefore reject a perfectly good compromise policy. Conversely, people may find it hard to infer which policy is implied by a given set of actual policy choices, when trying to find an acceptable alternative by working from concrete cases.

POLICY can help with both of these problems. Thus, it is possible to use the program to derive predicted judgments for a set of actual problems in order to test the implications of a given policy, as well as to use it to find the policy implied by a given set of actual choices.

The study by Flack and Summers (1971) provides an example of how POLICY might be used. Flack and Summers were concerned with conflicts over water resources planning. In their study, two Corps of Engineers employees made judgments about a series of hypothetical projects. Their policies were then analyzed, and the differences in policy causing conflict identified. After these differences had been displayed to the participants, they were asked to specify new policies by assigning a new set of weights to the factors describing each project. These weights were then used by POLICY to generate a set of "error free" judgments so that the participants could see the implications of their new policies. As it turned out, the new policies resulted in *more* conflict, thus confirming that people may actually have problems understanding the implications of a new policy. The participants were then asked to discuss the water resources problem until they could find a set of weights which they could agree upon. Such a policy was found eventually, and conflict was resolved.

Presumably, POLICY aided conflict resolution in the present case in two different ways. First, it allowed the persons in conflict to focus upon what really mattered, i.e., the underlying policy, rather than individual cases. Thus, POLICY served as an aid to communication. Second, it helped the persons realize what the consequences of their policies were, and this made it possible for them to select an adequate policy.

A similar application of POLICY is described by Steinmann, Smith, Jurdem and Hammond (1977). This case involved finding a policy for acquiring open space by the city council of Boulder. In this case, a number of factions with different policies were identified on the basis of an analysis of the judgments policies of the members of the city council. A compromise policy was then worked out and presented to the members of the council, together with information which made it possible to compare the decisions which they would make under this new policy to the decisions they would have made with their old policies. This information made clear that the new decisions would not differ all that much from the decisions made under the old policies, and the new policy was then adopted by the council.

Although the compromise policy was adopted by the council, it was used only for a few months. Observations during the period in which the policy was in use showed that the procedures which the board members had been taught were seldom used, and when they were used, they were often used in an inappropriate way. Thus, although the attempt to resolve conflict was successful, the success did not last. This suggests that if procedures of this kind are to be used, it is necessary to train the users better, and perhaps also to provide continuing advisory service (Steinmann et al., 1977).

5. SEPARATION OF FACT AND VALUE IN COGNITIVE CONFLICT

In 1974 the Denver police wanted a more effective bullet for their handguns, and selected a so-called " hollow point" bullet. This was vehemently opposed by various citizen groups, who saw this new kind of ammunition as a kind of "dum-dum bullet".

Hammond, Stewart, Adelman and Wascoe (1974; see also Hammond and Adelman, 1976) noted that the discussion about the bullets confused problems of value (what one wanted the bullet to do) and problems of fact (what a given bullet actually does), and that different persons should be involved in deciding the value issue and the fact issue. The value problem was a question for the citizens of Denver and their representatives, while the fact question should be settled by ballistics experts.

There was no direct way of resolving the issue, because the information available about various kinds of bullets did not speak directly to the concerns of the citizens. It was therefore necessary to clarify which aspects of bullets were of importance to the citizens and how bullets compared with respect to these aspects.

Agreement was quickly reached that three aspects of bullets were important: its "stopping power" (i.e., whether a person who had been hit by a given bullet would be likely to fire back at the police officer), the threat to bystanders (whether one could expect ricochettes, and so on), and the degree of injury caused by the bullet.

A number of hypothetical bullets which differed along these dimensions were then presented to members of the Denver city council, the mayor, representatives of the various interest groups, and members of the general public, who judged each bullet with respect to acceptability. Their judgments were then subjected to judgment analysis to provide information about the relative importance of the three characteristics of the bullets. The results showed that there were a number of factions, i.e., groups which differed with respect to their weighting schemes. The city council was, however, able to achieve a compromise and adopted a policy that gave equal weights to the three factors. Thus, it was possible to reach agreement on the value issue.

The next step involved the fact issue. Three ballistics experts were asked to make judgments about a number of available bullets with respect to the three dimensions: stopping power, lethality and threat to bystanders. The judgments of the ballistics experts were then used in computing the acceptability of each bullet by means of the policy adopted by the city council. From these acceptability scores, a bullet was finally selected.

This provides an example of how facts and values can first be separated, and then integrated again, so that it becomes possible to reach a decision also under conditions of manifest conflict. A similar analysis involving separation of facts and values in a different policy issue: that of which parcels of land the Boulder city council should acquire, is presented in Steinmann et al. (1977).

Despite the fact that both of these cases started out as cases of manifest conflict, it proved comparatively simple to reach agreement once facts and values had been separated. However, things are not always this simple, for in many cases we must expect not only that there is disagreement with respect to values, but also that there will be disagreement with respect to facts among the experts who are to provide these. We now turn to this problem.

6. WHEN EXPERTS DISAGREE

Recent debates about the introduction of new technologies have shown that it is not always possible to secure unanimous expert opinion. This is not surprising; problems of the magnitude and kind raised by the introduction of nuclear power, for example, are so complex in relation to the knowledge provided by scientific theories and data that no certain answer is possible. Moreover, the kinds of variables and factors that are relevant in a given decision problem may not necessarily be the kinds of variables studied in scientific research. Therefore, experts often can not compute the answers wanted by the decision makers, but have to rely upon their judgment. Consequently, we would expect to find disagreement, as indeed we do.

Differences among experts constitute a case of cognitive conflict. The question is whether such conflicts can be resolved and if so, by what means. Hammond, Anderson, Sutherland and Marvin (1982) present one successful case involving radiation experts.

The problem in this case arose because the Rocky Flats Monitoring Committee, which was appointed by the governor of Colorado to monitor risk of cancer from plutonium emissions from a federally owned industrial plant, could not secure agreement among its experts about the level of risk associated with different levels of emission. The situation had become increasingly difficult, with acrimonious disputes among the scientists which prevented the committee from fulfilling its task.

Hammond and his associates attacked this problem in three steps. First, judgment analyses were performed to uncover whether there were stable differences in judgment policies among scientists. The judgment task required each scientist to make judgments about cancer risk for a set of individuals from information about their exposure to various kinds of cancer agents. The results showed that there were indeed stable differences among scientists, but that these differences could not be attributed to the training of the scientists.

The second step involved an attempt to resolve the differences among the experts. The five experts with the most diverging judgment policies were chosen. They were asked to work out an analytical procedure for making judgments about risk. This required them to defend their choice of variables to be included in the prediction equation and the forms of the relation between each variable and the risk level on the basis of publicized scientific information. This procedure was successful. It was possible for the five diverging experts to agree upon which factors should be considered and what the relations between these factors and the level of risk should be.

In the third step, an explicit procedure for computing the actual level of risk for persons with different kinds of exposures was developed.

There are two important elements in this study. The first is the use of judgment analysis to demonstrate that there were differences in how the scientists thought about the problem of risk. The second involved finding a procedure that moved the scientists from the quasi-rational mode of making judgments, which had led into conflict, to a more analytical mode, where they could find agreement.

The first element involves the demonstration that the disagreement had a real cognitive basis, and that it thus had to be resolved by cognitive means. This was by no means clear at the outset. Presumably, the demonstration that there were such differences provided an important source of motivation for the scientists to take part in the attempt to find such a cognitive means. The second element was the particular cognitive means that was employed.

The first element should have considerable generality; a judgment analysis can always be performed. The second element may be less general. In the case of radiation, there exists a considerable body of data that can be used to work out a policy in the analytical mode. For many problems involving new technology, this may not be the case. However, discovering that it is indeed impossible to work out an agreed upon policy in the analytical mode may have a sobering effect, and reduce unnecessary conflict.

7. FINDING A POLICY FOR A COMMUNITY

Is a decision aid such as POLICY applicable only to so-called "decision makers", or can it be used also by the general public to express their underlying policies about important issues as an input to the decision process? A study by Rohrbaugh (1975) addresses this issue.

Rohrbaugh placed computer terminals in two grocery stores in a small town in Colorado, and invited people coming to the store to express their views about possible future developments of their community. They did this simply by making judgments about the acceptability of a number of alternative possible futures for their community, futures that were described in terms of five different factors. Their judgments were then analyzed and factions which differed with respect to their policies were identified.

The results indicated that the general citizen found POLICY easy to use. Four different factions were identified, but a compromise policy close to all these factions was relatively easy to find. This suggests that it may be possible to use the POLICY method to identify the policies in a community, and to try out a workable compromise policy for that community.

8. CONCLUSIONS

The results presented in this paper support the view that cognitive factors are sufficient to cause conflict and to prevent conflict resolution, and that conflicts caused by cognitive factors can be resolved if the appropriate cognitive aids are provided. These aids help to resolve conflict by improving communication, and by allowing people to try out different policies under full cognitive control so that the consequences of different kinds of policy change can be investigated.

Do these methods in any way help to resolve the cognitive conflicts caused by new technology? So far, there have been no actual analyses of such conflicts, but as was discussed in the introduction to this paper, there are good *a priori* reasons for believing that such conflicts have strong cognitive elements. Thus, there are reasons to believe that the kinds of conflict analyses proposed by social judgment theorists will be relevant to these conflicts. As for the question of whether the *methods* developed in SJT will help to resolve the conflicts, there are grounds both for pessimism and for optimism.

As for the pessimistic side, it is clear that the cognitive aids developed by social judgment theorists can not resolve genuine uncertainty about the future. Policy analysis is no substitute for knowledge, and where no agreement about the facts can be reached because there is not enough information, conflicts can not be resolved. Many of the issues raised by the introduction of new technology may very well have this character, and where this is true, we can not hope to resolve the conflict, we can only hope to unearth its real causes.

However, that is no small thing, and it may prevent prolonged and needless dispute.

As for the optimistic side, the methods developed by social judgment theorists can help to set the stage for a more fruitful discussion by identifying the real reasons for the conflict, and by allowing the parties to focus upon the real issues, i.e., the underlying policy, to assess the extent to which different options are consistent with this policy, and to separate questions of fact from questions of value.

Thus, it seems that SJT methods hold promise for helping to resolve some of the problems caused by the introduction of new technology. It is certainly time that we try these methods, and assess their value in that context.

REFERENCES

Balke, W.M., Hammond, K.R. and Meyer, G.D. (1973). An alternative approach to labor-management relations. *Administrative Science Quarterly, 18*, 311-327.

Brehmer, B. (1974). Policy conflict, policy consistency, and interpersonal understanding. *Scandinavian Journal of Psychology, 15*, 273-276.

Brehmer, B. (1976). Social judgment theory and the analysis of interpersonal conflict. *Psychological Bulletin, 83*, 983-1003.

Brehmer, B. (1984). The role of judgment in small group conflict and decision making. In: G. Stephenson and J. Davis (Eds.), *Progress in applied social psychology. Vol. 2* (pp. 163-184). Chichester: Wiley.

Brehmer, B. (1987). Social judgment theory and forecasting. In: G. Wright and P. Ayton (Eds.), *Judgmental forecasting* (pp. 199-214). Chichester: Wiley.

Brehmer, B., Hagafors, R. and Johansson, R. (1980). Cognitive skills in judgment: Subjects' ability to use information about weights, functional forms, and organizing principles. *Organizational Behavior and Human Performance, 26*, 373-385.

Chen, K. and Mathes, J.C. (this volume). Value oriented social decision analysis: a communication tool for public decision making on technological projects.

Cook, R.L. and Hammond, K.R. (1982). Interpersonal learning and interpersonal conflict reduction in decision making groups. In: R.A. Guzzo (Ed.), *Improving group decision making in organizations* (pp. 13-40). New York: Academic Press.

Flack, J.E. and Summers, D.A. (1971). Computer aided conflict resolution in water resource planning: An illustration. *Water Resources Research, 7,* 1410-1414.

Hammond, K.R. (1982). *Principles of organization in intuitive and analytical cognition.* (Report No. 226). Boulder: University of Colorado, Center for Research on Judgment and Policy.

Hammond, K.R. and Adelman, L. (1976). Science, values, and human judgment. *Science, 194,* 389-396.

Hammond, K.R., Anderson, B.F., Sutherland, J., and Marvin, B. (1982). *Improving scientists' judgments of risk.* (Report No. 239). Boulder: University of Colorado, Center for Research on Judgment and Policy.

Hammond, K.R. and Brehmer, B. (1973). Quasi-rationality and distrust: Implications for international conflict. In: L. Rappoport and D. Summers (Eds.), *Human judgment and social interaction* (pp. 338-391). New York: Holt, Rinehart and, Winston.

Hammond, K.R., Hamm, R.M., Grassia, J. and Pearson, T. (1984). *The relative efficacy of intuitive and analytical cognition.* (Report No. 252). Boulder: University of Colorado, Center for Research on Judgment and Policy.

Hammond, K.R., Stewart, T.R., Adelman, L. and Wascoe, N. (1974). *Report to the Denver city council and mayor regarding the choice of handgun ammunition for the Denver police department.* (Report No 179). Boulder: University of Colorado, Center for Research on Judgment and Policy.

Hammond, K.R., Stewart, T., Brehmer, B., and Steinmann, D.O. (1975). Social judgment theory. In M. Kaplan and S. Schwartz (Eds.), *Human judgment and decision processes* (pp. 271-312). New York: Academic Press.

Hammond, K.R. Rohrbaugh, J., Mumpower, J., and Adelman, L. (1977). Social judgment theory: Applications in policy formation. In: M. Kaplan and S. Schwartz (Eds.), *Human judgment and decision processes in applied settings* (pp. 1-27). New York: Academic Press.

Libby, R. and Lewis, B.L. (1982). Human information processing research in accounting: The state of the art in 1982. *Accounting, Organizations, and Society, 7,* 231-285.

Phillips, L.D. (this volume). Requisite decision modelling for technological projects.

Rohrbaugh, J. (1975). Cognitive maps: Describing the policy ecology of a community. *Great Plains-Rocky Mountains Geographical Journal,*

6, 64-73.

Slovic, P. and Lichtenstein, S. (1971). Comparison of Bayesian an regression approaches to the study of information processing in judgment. *Organizational Behavior and Human Performance, 6,* 649-744.

Steinmann, D.O., Smith, T.H., Jurdem, L.G. and Hammond, K.R. (1977). Application and evaluation of social judgment theory in policy formation: An example. *Journal of Applied Behavioral Science, 13,* 69-88.

Stewart, T.R. and Carter, J.E. (1973). *POLICY: An interactive computer program for externalizing, executing and refining judgmental policy.* (Report No. 159). Boulder: University of Colorado, Center for Research on Judgment and Policy.

Vedung, E. (1979). Kernkraften ger ny blockbildning i politiken. *Tvärsnitt, 1,* 42-47.

NOTES

1. Work on the paper was supported by a grant from the Swedish Council for Research in the Humanities and Social Sciences.

APPROACHES TOWARDS CONFLICT RESOLUTION IN DECISION PROCESSES[1]

ANNA VARI

Hungarian Institute for Public Opinion Research, Budapest

1. INTRODUCTION

Strategic decision problems are complex, ill-structured, inter- or intra-organizational problems, characterized by a number of stakeholders looking at the problems from different perspectives (Linstone, 1984). It has been revealed in several studies that significant differences in the views of the stakeholders may exist, concerning: (i) the problem to be solved, (ii) objectives and the hierarchy of goals, (iii) past and future states of the environment, and (iv) possible alternative actions, their consequences and associated value; (Checkland, 1981; Mason and Mitroff, 1981; Kleindorfer and Yoon, 1984).

These differences in the stakeholders' views may lead to conflicts during the decision process. Conflicts can be handled in many different ways, e.g., by forcing a single solution, by mediation, via legal procedures. In the present paper the possibilities of conflict management by using the tools and methods of decision support are investigated. First, we give a classification of conflicts arising in strategic decisions as well as possible sources of conflicts. This is followed by a short description of alternative approaches and methods applicable to the resolution of different conflicts. Finally, guidelines are given for the application of these approaches and methods. The outlined principles are illustrated in two case-histories.

CASE 1: *Starting a new venture - a textile company*
Five young managers in the textile industry decided to leave their previous jobs because they were unsatisfied with their status at that company. They met with the president of a Sports Club who wanted to extend the Club's income with a new profit-oriented business.They began negotiations to establish a new textile company in the frame of the Sports Club. These negotiations were supported by a decision analysis process in which the objectives of both parties, the feasible

79

Ch. Vlek and G. Cvetkovich (eds.), Social Decision Methodology for Technological Projects, 79–94.
© *1989 by Kluwer Academic Publishers.*

strategies, including the business profile, scale, and organization, were defined and evaluated.

CASE 2: *Siting a hazardous waste incinerator*
It was decided by three Hungarian pharmaceutical companies to establish a joint venture for the elimination of their hazardous wastes. They planned to site the incinerator in a small town, 700 meters from a residential area. According to the plans, half of the capacity of the incinerator was supposed to be adequate for eliminating the waste produced in that town. The remaining half would be used for incinerating the waste of units of the pharmaceutical companies in other towns. When the news of the siting plan leaked out, the residents of the town which had an already overpolluted air condition started to protest. The safety of the incineration technology was questioned and an alternative technology and/or siting in the countryside was suggested. As a consequence of the protest, negotiations were opened between proponents of the incinerator and the residents. The main issues of negotiation included the technology, the location of the siting, the guarantees about the safety of the incinerator, and compensations. During the negotiations, no solution satisfying all parties could be reached.

No systematic supporting procedures were used for the negotiations. The reasons for the failure of the negotiations were revealed by interviews after the decision was taken (Vári, Vecsenyi and Paprika, 1986; Faragó, Oldfield and Vári, 1988) and are discussed later.

2. CLASSIFICATION OF CONFLICTS DURING DECISION MAKING

Several attempts have been made to design taxonomies for classifying conflicts. A starting point of most attempts is to distinguish between disagreements in opinions and beliefs ("facts") and disagreements in values and concerns ("values"). This classification was further refined by Von Winterfeldt and Edwards (1984) who developed a hierarchical system of conflicts over technology. The main deficiency of their classification is that it does not differentiate between the disagreements manifested in discussions and the sources of such conflicts.

Humphreys and Berkeley (1984) have classified conflicts according to a five-level hierarchy of problem-structuring operations, as follows:

level 5:   Exploring problem boundaries

level 4:   Identifying and linking variants of
             judgemental frames   }   Developing
level 3:   Developing structure within a   problem
             particular frame   structure

Level 2:   Making conditional assessments   Making
           of data   }   data
Level 1:   Making "best assessments" of data   assessments

Humphreys and Berkeley (1984) pointed out that conflicts can arise at each level and they can be resolved only in a top-down manner, e.g., there is no use in reconciling data assessments if different structures are used by stakeholders for representing the same problem.

Mumpower (1987; see also Mumpower, Schuman and Zumbolo, 1988) classified sources of conflict in terms of the stages of the negotiation process where they can arise. He suggests three stages of negotiation: (1) independent evaluation of alternative solutions by each party, (2) further evaluation taking into account additional relevant factors, and (3) reaching agreement on a mutually acceptable solution. Taking a closer look at the individual stages, we may divide the decision making and negotiation process into the following phases:

phase 1:   Problem recognition   }   A. Problem
phase 2:   Problem definition   formulation

phase 3:   Generation of alternative solutions   B. Scenario
phase 4:   Analysis of the consequences of   }   analysis
             solutions

phase 5:   Evaluation of alternative solutions   C. Evaluation
phase 6:   Selection of the best (acceptable) }   and choice
             solution

We found it promising to combine Humphreys and Berkeley's (1984) operation-oriented frame with the process-oriented one above for classifying the different types of conflict, for the following reasons. First, it helps to define the hierarchy and sequence of the conflict resolution methods to be applied. Secondly, it makes it possible to clearly differentiate between the disagreements manifested in the different phases and levels of the decision making process on the one hand, and the underlying sources of conflict on the other.

We found that the three-level, three-stage frame displayed in Table 1 is complex enough for classifying the different types of conflict with a view on the applicable conflict resolution methods.

Table 1: The phases and levels of the decision making process.

LEVELS OF OPE- RATIONS	PHASES IN THE PROCESS	PROBLEM FORMULATION (A)	SCENARIO ANALYSIS (B)	EVALUATION AND CHOICE (C)
EXPLORING PROBLEM BOUNDARIES		A.Defining problem boundaries	-	-
DEVELOPING PROBLEM STRUCTURE		-	B1.Developing structure of scenarios	C1.Developing evaluation structure
MAKING ASSESSMENTS		-	B2.Assessing scenario data (probabilities)	C2.Assessing evaluationdata (utilities, trade-offs)

In the above frame we have identified the following five types of conflict: (A) conflicts about problem boundaries, (Bl) conflicts about the structure of scenarios, (B2) conflicts about scenario data, (C1) conflicts about the evaluation structure, and (C2) conflicts about evaluation data. These conflicts will be outlined by analysing the case-histories.

2.1 Conflicts about problem boundaries (A)

This type of conflict appears in the disagreements about the problem context, about the objectives, about the supposedly exogenous and controllable factors, and about what questions are to be answered.

In the case of *the textile company*, the objective of the president of the Sports Club was to gain financial support while the young managers' intention was to establish a dynamically developing, high-prestige organization where they could actualize themselves. While the president's decision was influenced mainly by economic and political considerations (e.g., the attitude of the leaders of the town, public opinion), the decision of the managers was rather affected by individual and organizational factors.

The interviews on *the hazardous waste incinerator* disclosed that the problem for the proponents of the incinerator primarily had technical and financial implications, while for the residents the health and ecological consequences were of most relevance. For the proponents it seemed to be indispensable to establish the incinerator and in their view only the location was questionable. The residents, however questioned altogether the necessity of the incinerator itself and they proposed alternative waste-neutralizing technologies.

2.2 *Conflicts about the structure of scenarios (B1)*

The second type of conflict is related to the differences in views on the decision alternatives and their consequences, i.e., on the structure of scenarios. Investigations by Lathrop and Linnerooth (1983) have revealed that the model structure describing the future scenarios and the factors worthy of consideration may essentially differ for the conflicting parties in cases of multi-stakeholder decision problems (e.g., decisions involving environmental risks).

This seems to be confirmed by our studies on the siting of *the hazardous waste incinerator*. These studies show that: (i) more alternative solutions on technology and location of siting are considered feasible by the residents than by the proponents, and (ii) that the residents more clearly imagine the short- and long-term risks of the siting and the potential accidents following a more complex scenario than the proponents; the majority of the proponents even refuse to reveal risk scenarios.

Feasibility testing in the case of *the textile company* showed that there were no marked differences in opinions between the president and the managers as to the structure of future scenarios. They all envisaged that similar actions would be taken by the same organizations (competitors, local government, banks, etc.); only the probability estimates concerning these actions were different.

2.3 Conflicts about scenario data (B2)

One of the most frequent types of conflict concerns the data included in the scenarios. This may appear if the parties disagree about data, statistics or facts (see Von Winterfeldt and Edwards, 1984). Differences may be even more frequent with respect to future forecasts within scenarios, e.g., cost estimates or probabilities.

The feasibility testing in *the textile company* case revealed significant differences in views concerning the probability of the support or opposition from the organizations having a key role in the development of the new company.

Concerning *the hazardous waste incinerator*, the contradictions appeared in the estimates of investment costs, the measure of air pollution, and the probabilities of breakdown and of transportation accidents.

2.4 Conflicts about the evaluation structure (C1)

This type of conflict results from the use of the different evaluation models, (e.g., additive vs. non-additive utility models) and different assessment criteria, used by the different stakeholders.

In the case of *the textile company*, the alternative action strategies were evaluated by the president of the Sports Club mainly in terms of possible income, while the managers were focussed on criteria of individual and organizational development.

The differences in the preference structure were even more striking in the case of *the hazardous waste incinerator*. For the sponsors, the minimization of investment costs was the most important criterion, while for the residents short- and long-term safety got the highest importance rating.

The differences in the evaluation models may appear not only in the different character of the criteria but also in different interpretations of the same criteria. Concerning risk debates, the phenomenon of different interpretations is rather frequent. For example, industrial experts characterize the risks by the probability of hazard or by the expected annual fatality rate. As a consequence, they often calculate low risks. On the other hand, opponents estimate the consequences of the largest plausible accidents and come up with large fatality risks (see Von Winterfeldt and Edwards, 1984).

2.5 Conflicts about evaluation data (C2)

One group of this type of conflict is related to differences in utility functions, while the other one is related to different weightings of the evaluation criteria. With respect to the first type of conflict Mumpower (1987) points out that negotiating parties may value outcomes in different, sometimes opposite ways. Conventional theories describe these situations in terms of zero-sum, differential games, but as Mumpower points out, higher or lower negative correlations of utility functions across some criteria can occur in various other kinds of situation.

Another group of conflicts at the level of evaluation data concerns the disagreement on trade-offs between conflicting criteria. The criteria having a cost-risk-benefit character can be differently weighted. In some cases this may lead to totally different assessments. However, this is not necessarily so. For example, in case of a positive correlation of assessments by criteria various weights may produce the same results (Mumpower, 1987).

In the case of *the textile company*, of the alternative strategies with respect to economic criteria, those were the better which met the criteria of technical level, development and self-actualization. It is not surprising, therefore, that a strategy could be found which proved to be the best within a wide range of changes of criterion weights.

In the case of the siting of *the hazardous waste incinerator* the criteria of safety and economic efficiency seemed to be in contradiction. Nevertheless, it now appears that a solution could have been found which would fulfil both requirements (i.e., an incinerator far from any residential area, which, however, would be of a large capacity by producing energy for the community).

2.6 Sources of conflicts

In the above taxonomy we differentiated between conflicts appearing in different phases and levels of the decision process, but did not consider the sources of conflicts. Analyses of conflicts may refer to several potential sources like, e.g., differences in cultural and educational background, differences in available information and its interpretation, conflicting interests, value differences, lack of confidence, etc. Evidently these factors are not mutually exclusive but they may appear with strong interdependencies. From the point of view of choosing an adequate conflict resolution method the actual sources of conflict are of crucial importance.

3. APPROACHES TOWARDS CONFLICT RESOLUTION

As the starting point for presenting approaches towards the resolution of conflicts occurring at various levels and phases of the decision making process, we consider the taxonomy of Kilman and Thomas (1974; see also Thomas, 1976). This is shown in Figure 1.

Figure 1: Ways of reacting to conflict; the numbers in parentheses represent (potential) gains for parties A (first) and B.

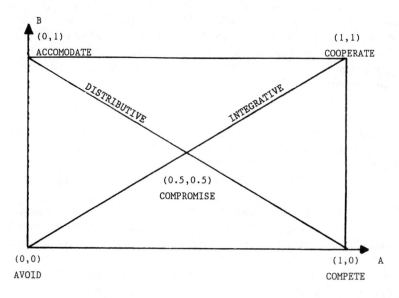

Figure 1 shows the basic ways of responding to a conflict situation, in the case of two parties A and B. The point (0,0) represents the case where both parties decide to avoid the conflict. Competition, on the other hand, is the case where one of the parties decides to get all it can at the expense of the other party. Accommodation, at the other extreme, represents A giving in completely to B's wishes. Competition and accommodation are opposites, represented, respectively, by the points (1,0) and (0,1). The line connecting these two points is the *distributive* dimension; it involves the points representing different compromises.

Points in the distributive dimension represent the world as a zero-sum game, but there is another dimension which does not represent the world like this. Points in the *integrative* dimension, e.g., point (1,1), represent cases where both parties can increase their gains.

Based on the above taxonomy we suggest three basic approaches towards conflict resolution:

(i) The *reconciling approach*, aimed at finding a solution satisfying each party without trying to remove the disagreements at all levels and in all phases of decision making. It corresponds to the point (0,0) of conflict avoidance in Figure 1. This approach is appropriate in cases where conflicts arising from opposing interests, different values and mistrust are not significant.

(ii) The *bargaining approach*, aimed at finding a compromise acceptable for each party. It corresponds to the set of points on the distributive dimension (line (1,0)-(0,1)). This approach can be suggested in cases with highly opposing interests, strong value differences and mistrust.

(iii) The *confronting approach*, aimed at finding a creative solution by removing the disagreements at all levels and in all phases of decision making, via direct confrontation of the different opinions. It corresponds to the set of points on the integrative dimension (line (0,0)-(1,1)) in Figure 1. The main application area of this approach consists of situations where the sources of conflicts are uncertain or where the former two approaches have already failed.

4. APPLICATION OF CONFLICT RESOLUTION APPROACHES

In the following the methods used within the above approaches are briefly outlined.

4.1 The reconciling approach

The essence of this approach is that it does not confront the differences in opinions but it attempts to unify the different problem representations of the stakeholders mainly by group decision support. In case of type A conflicts (see Table 1) this means that the problem is put into as wide a context as is required to cover all individual problem contexts. This procedure can be rendered more effective by demonstrating and uniting the individual problem representations by means of inference diagrams (see Vári et al., 1986).

Regarding type Bl and Cl conflicts, models can be constructed which include all the options, possible consequences, outcomes and the evaluation criteria considered by all the parties. This is called by Phillips (1982, 1984, see also this volume) requisite modelling where the model "is requisite in the sense that everything required to solve the problem is either included in the model or can be simulated in it. To develop a requisite model, it is necessary to involve all those who are in some way responsible for aspects of the decision in the development of the requisite model."

In the case of type B2 and C2 conflicts, sensitivity analysis by changing the inputs to the model, plays a key role "in helping to resolve disagreements about the implications of differing assumptions on the decision itself, and in showing the extent of disagreement about assessments and judgements that can be tolerated for a given decision" (Phillips, 1982).

The reconciling approach mainly relies on decision-analytical models (decision trees, multi-attribute utility models, etc.) for providing a framework for the iterative development of a coherent representation of the problem. The models are developed in group sessions assisted by facilitators often using group techniques (brain-storming, nominal group technique, etc.). Such analyses are mostly conducted in decision-analysis projects or at decision conferences.

As has already been mentioned, this approach can be used only if there are no basic differences in interests and values and if participants are ready to provide each other with the information relevant to joint decisions. These are, first of all, the *intra*-organizational problems. However, such situations may occur in *inter*-organizational decisions, too (e.g., in formulating a joint business policy of several companies against their competitors).

A largely reconciling approach was employed in the case of *the textile company*. A joint meeting was organized with the participation of the president of the Sports Club and the young managers, where both parties defined their objectives and constraints, and found a problem representation uniting their perspectives.

The alternative action strategies, such as the variants of the business profile, scale and organization, were generated in goal-driven scenarios by developing a requisite model in group sessions. However, a confronting approach was used during the feasibility testing of the strategies since disagreements about implementability were significant. The Strategic Assumption Surfacing and Testing (SAST) method (Mason and Mitroff, 1981; see also Section 4.3 below) was applied to confront assumptions about the reactions of the possible stakeholders (competitors, local government, bank, etc.) concerning the various strategies. The SAST analysis revealed the differences in the underlying assumptions and the uncertainties calling for further information.

After the guided information search the group reached an agreement on the assumptions as well as on the feasibility of the alternative strategies. The feasible strategies were evaluated by means of a multi-attribute utility model. Since this model incorporated all the criteria suggested by the participants, the model can be regarded as 'requisite'. For analyzing the effects of disagreements in evaluation on ranking, sensitivity analysis was carried out. This revealed that the rank-order of strategies was not sensitive to these differences, so the analysis was concluded by finding a solution satisfying both parties.

4.2 The bargaining approach

Here, a representation and a solution of the problem is reached which is the result of a deliberate compromise between the parties. It assumes that the views of the parties may come closer to each other, or where they are not, a reasonable compromise can be arranged by using mathematical models. With regards to the type A conflicts, a problem representation uniting certain elements of the various perspectives is created, where the range of interpretation is the result of the bargaining process between the parties.

With B1 and C1 conflicts, elements of the scenarios as well as of the evaluation model to be considered may also be the outcome of bargaining. For compromising on B2 and C2 type conflicts, various statistical procedures are known (e.g., calculation of expected values, standard deviations, coefficients of concordance, etc.). The other type of methods to be considered for approaching especially C2 type conflicts are the traditional negotiation and bargaining procedures (Fisher and Ury, 1983) and the negotiation support models (Nash, 1950; Raiffa, 1982). Taking into consideration the different preference structures of the parties, these models identify solutions that simultaneously maximize joint desirability.

The application of the bargaining approach requires the participation of mediators who intend to promote communication between the parties, aiming at reaching a compromise on the basis of information they have received from each party. For instance, in contrast to a decision-conference-like approach, the bargaining model and the related data (e.g., the utility functions and the trade-offs of the opposing parties) may not be familiar to all parties, only the optimum suggested by the model acts as a basis for further negotiations.

Adversarial situations with considerable opposing interests represent the main field of application for the bargaining approach. These are primarily inter-organizational problems although such situations may also arise within one organization (e.g., in employer-employee debates).

A bargaining approach was used by the proponents of *the hazardous waste incinerator* for eliminating the siting conflict. For example, while the residents suggested siting in the countryside, the proponents were willing to change the location of the siting only by 150 meters. In return for the siting a compensation was offered. A bargaining process was started, since the residents wanted to use the financial compensation for cleaning up the air of the town (instalment of a gas pine-line, investments for environmental protection equipments in the existing plants, etc.), but the sum involved did not cover the intended expenses. Finally, the proponents only slightly adapted their original proposal in response to the residents' demands, and the resulting decision, therefore, did not satisfy the residents.

4.3 The confronting approach

In contrast to the former two approaches, this one focusses on the differences in problem representations. It is based on the assumptions: (i) that the revealing and confrontation of the different opinions may help in exploring the sources of conflicts, and (ii) that the exploration of the sources of conflict may facilitate the finding of creative solutions.

For studying type A conflicts (see Section 2.1), text-analytical procedures are suggested (Gallhofer, Saris and Melman, 1987; Vári, 1988, in press). These attempt to reveal the assumptions, cognitive structures and motivations underlying the different problem formulations based on the verbal manifestations (interviews, documents) of the stakeholders.

For handling B1 and B2 type scenario conflicts the SAST (see Section 4.1) planning process developed by Mason and Mitroff (1981) can be used. As they point out, the process "is based on the premise that the best judgement on the assumptions necessary to deal with a complex problem is rendered in the context of opposition."

In the SAST process the groups having different perspectives on the action strategy are engaged in an effort to debate their underlying assumptions regarding the plausible future scenarios, in order to synthesize the assumptions and to establish guidelines for the required actions to take the decision (gathering of further information, modelling, etc.).

The SAST procedure is a special version of argumentation analysis for managing scenario conflicts. According to Yoon (1986) argumentation analysis can also be used for handling C1 and C2 type evaluation conflicts. This analysis helps in differentiating between conflicts resulting from different knowledge, opposing interests, different social values, etc.

The main benefit of the confronting approach is that it helps in finding the conflict management method most appropriate for the given situation. In case of knowledge differences, gathering of new data or exchanging the available information might be a solution. In case of opposing interests, the use of a bargaining process could be useful. With value differences requisite modelling and sensitivity analysis may come in handy.

The confronting approach is not contradictory but rather complementary to the former two approaches. It was found that in some cases it indicates when the other two approaches should be selected. At the same time, confronting methods are recommended in certain phases of reconciling type decision analysis projects or of bargaining processes, if a deeper understanding of the sources of conflict is necessary.

A confronting approach was adopted in a retrospective analysis of *the hazardous waste incinerator* case. Analyzing arguments and counter-arguments revealed: (i) that the incinerator was not judged to be safe enough by the residents; they refused to accept the arguments of the proponents based on statistics and low-probability estimates, since the residents interpreted risk in terms of the severity of the negative consequences of possible accidents, (ii) that safety played an exclusive role in the preference structure of the residents who did not consider financial support compensatory for health risks, and (iii) that the residents did neither believe in the promised measures for cleaning up the air nor in the reassuring information provided by the proponents.

The analysis primarily disclosed the conflicting values and the lack of mutual trust between the parties involved. The lack of information was less important. This result accounts for the fact that actions towards informing the residents and the proposition of compensation had failed. Instead, efforts strengthening mutual trust and the selection of a siting location judged safe enough by the residents, would have been needed.

5. CONCLUSIONS

In this paper a new taxonomy for classifying conflicts during strategic decision making is proposed. A distinction is made among conflicts arising at different operational levels and in different phases of a decision process. Together with a separate consideration of the (potential) *sources* of conflicts, this may be useful in formulating guidelines for the application of different approaches and methods of conflict resolution. In our experience we found that especially in the case of *intra*-organizational problems most of the conflicts can be treated with the help of existing decision support methods.

There are, however, conflicts - especially in *inter*-organizational problem solving - whose solution requires new institutional mechanisms, measures for strengthening trust and an intensive training and learning process (see, e.g., Chen and Mathes, Dienel, and Hisschemöller and Midden, this volume). The above-presented analyses and approaches may stimulate the development of proposals which may well serve to significantly improve upon the traditional process of "muddling through" (Lindblom, 1964).

REFERENCES

Checkland, P. (1981). *Systems thinking, systems practice.* New York: Wiley.

Chen, K. and Mathes, J.C. (this volume). Value oriented social decion analysis: a communication tool for public decision making on technological projects.

Dienel, P.C. (this volume). Contributing to social decision methodology: citizen reports on technological projects.

Faragó, K., Oldfield, A. and Vári, A. (1988). Conflicting perspectives in multi-stakeholder problems: a comparative study. In: J. Abbou (Ed.), *Hazardous waste, detection, control, treatment.* Amsterdam: Elsevier-North-Holland.

Fisher, R. and Ury, W. (1983). *Getting to yes.* New York: Penguin Books.

Gallhofer, I.N., Saris, W.E. and Melman, M. (1987). *Different text analytical procedures for the study of decision making.* Amsterdam: Sociometric Research Foundation.

Hisschemöller, M. and Midden, C.J.H. (this volume). Technological risk, policy theories and public perception in connection with the siting of hazardous facilities.

Humphreys, P.C. and Berkeley, D. (1984). *How to avoid misjudging judgement.* Decision Analysis Unit, London School of Economics and Political Science. London.

Kilman, R.H., Thomas, K.W. (1974). *Four perspectives on conflict management*, Working Paper No. 86, Graduate School of Business, University of Pittsburgh.

Kleindorfer, P.R. and Yoon, T.H. (1984). *Toward a theory of strategic problem formulation.* Paper presented at the Fourth Annual Strategic Management Conference, Philadelphia, Pennsylvania.

Lathrop, J. and Linnerooth, J. (1983). The role of risk assessment in a political decision process. In: P.C. Humphreys, O. Svenson and A. Vári (Eds.), *Analysing and aiding decision processes* (pp. 39-68). Amsterdam: North-Holland, and Budapest: Akademiai Kiadó.

Lindblom, C.E. (1964). The science of "muddling through". In: W.J. Gore and J.W. Dyson (Eds.), *The making of decisions.* New York: The Free Press.

Linstone, H.A. (1984). *Multiple perspectives for decision making.* Amsterdam: North Holland.

Mason, R.O. and Mitroff, I.I. (1981). *Challenging strategic planning assumptions.* New York: Wiley.

Mumpower, J. (1987). *A theory of negotiation and mediation.* Unpublished manuscript. State University of New York, Department of Public Policy, Albany.

Mumpower,J., Schuman, S.P. and Zumbolo, A. (1988). Analytical mediation: an application for collective bargaining. In: R.M. Lee, A.M. McCosh and P. Migliarese (Eds.), *Organizational decision support systems* (pp. 61-73). Amsterdam: Elsevier Science Publishers.

Nash, J.F. (1950). The bargaining problem. *Econometrica, 18,* 155-162.

Phillips, L.D. (1982). Requisite decision modelling: a case study. *Journal of the Operational Research Society, 33,* 303-311.

Phillips, L.D. (1984). A theory of requisite decision models. *Acta Psychologica, 56,* 29-48.

Phillips, L.D. (this volume). Requisite decision modelling for technological projects.

Raiffa, H. (1982). *The art and science of negotiation.* Cambridge (Mass.): Belknap/Harvard University Press.

Thomas, K.W. (1976). Conflict and conflict management. In: M.D. Dunnette (Ed.), *Handbook of industrial and organizational psychology.* Chigago: Rand McNally.

Vári, A. (1988). *Argumatics: a text analysis procedure for supporting problem formulation.* Manuscript. Hungarian Institute for Public Opinion Research, Budapest.

Vári, A., Vecsenyi, J. and Paprika, Z. (1986). Supporting problem structuring in high level decisions. The case of the siting of a hazardous waste incinerator. In: B. Brehmer, H. Jungermann, P.F. Lourens and G. Sevón (Eds.), *New directions in research on decision making* (pp. 317-332). Amsterdam: North-Holland.

Von Winterfeldt, D. and Edwards, W. (1984). Patterns of conflicts about risky technologies. *Risk Analysis, 4(1),* 55-68.

Yoon, T.H. (1986). *Argumatics: a prescriptive approach for strategy formulation.* Manuscript, the Wharton School, University of Pennsylvania, Philadelphia.

NOTES

1. The text of this chapter has been revised, retitled and reprinted from R.M. Lee, A.M. McCosh and P. Migliarese (Eds.).(1988): *Organizational decision support systems*, pp. 87-100, with kind permission of the author and of Elsevier Science Publishers, Amsterdam.

REQUISITE DECISION MODELLING
FOR TECHNOLOGICAL PROJECTS

LAWRENCE D. PHILLIPS

Decision Analysis Unit
London School of Economics and Political Science

1. INTRODUCTION

Taking decisions about technological projects whose consequences will be experienced by different groups of people raises a host of difficult problems. Some of these are solved by applying some form of social decision analysis (Howard, 1975), which is the application of decision theory to problems in which there are multiple stakeholders. This paper is concerned with one type of social decision analysis that is called decision conferencing. This new approach to group decision making can contribute to the complex planning and decision making processes that accompany large-scale technological projects.

2. DECISION CONFERENCING

Decision conferencing is an intensive two-day problem-solving session attended by a group of people who are concerned about some complex issue facing an organisation. A unique feature of this approach is the creation, on-the-spot, of a computer-based model which incorporates the differing perspectives of the participants in the group. By examining the implications of the model, then changing it and trying out different assumptions, participants develop a shared understanding of the problem and reach agreement about what to do next.

The group is aided by at least two people from outside the organisation, a facilitator and a decision analyst, who are experienced in working with groups. The facilitator helps the participants to structure their discussion, think creatively and imaginatively about the problem, identify the issues, model the problem and interpret the results. The analyst attends to the computer modelling and helps the facilitator.

Decision conferencing draws on experience and research from three

95

Ch. Vlek and G. Cvetkovich (eds.), Social Decision Methodology for Technological Projects, 95–110.
© *1989 by Kluwer Academic Publishers.*

disciplines: decision theory, group processes and information technology. Decision theory (French, 1987) contributes to the development of the model, ensuring its internal consistency, so that subsequent changes to one part of the model do not require alterations to other parts that are considered satisfactory.

Research on group processes has identified conditions and situations that increase the ability of groups to solve problems effectively (Low and Bridger, 1979). Knowledge of small group functioning is used by the facilitator at the decision conference to help the group achieve its goals.

Finally, information technology in the form of a computer, computer programs and a tele-projection system, allows the model created by the group to be implemented on the spot, and provides the means for immediately showing the results. Thus, participants can try out different judgements, modifying their views or the model until a satisfactory representation of the problem is obtained. De Hoog, Breuker and Van Dijk (this volume) discuss various possibilities for computer support of (group) decision processes.

Although every decision conference is different, most are characterised by several stages that can be distinguished. After an initial introduction by the facilitator, the group is asked to discuss the issues and concerns that are to be the subject of the conference. An attempt is made to formulate the nature of the problem. Does the group wish to reconsider strategy, or is a fundamental change of direction required? Perhaps budget items or projects need to be prioritised. Evaluating alternative plans, ventures, systems, bids or projects may be required, especially if objectives conflict.

Once the nature of the problem has been formulated, the facilitator chooses a generic structural form for representing the problem, and the group begins to provide the content that is used in constructing the model. This is usually a simple, though not simplistic, representation of the group's thinking about the problem. Both data and subjective judgements are then added to the model, and the computer output is projected onto a large screen so all participants can see the results.

These initial results are rarely accepted by the group. Modifications are suggested by participants, and different judgements are tested.

Many sensitivity analyses are carried out; gradually, intuitions change and sharpen as the model goes through successive stages. Eventually this process of change stabilises, the model has served its purpose, and the group turns to summarising the key issues and conclusions. An action plan is created so that when participants return to work the next day, they can begin to implement the solution.

3. A CASE STUDY

To show how decision conferencing works, a case study will be presented. The example shows how some of the problems mentioned by other authors (Bezembinder, Vári, Wilke, all in this volume) can be reduced, if not entirely solved. These problems include the use of a fair rule that works in the collective best interest of all parties and that deals with the question of equality of outcomes, the generation of utilities that can be compared, the accommodation of objectives that may conflict, the provision of a process that enables people to understand others' points of view, and the social dilemmas problem.

The model developed in this case study is a specific realisation of a general class of resource allocation models that were developed at Decisions and Designs, Inc., of McLean, Virginia, mainly by Dr. Cameron Peterson, and implemented in a computer program, EQUITY, by Dr. Scott Barclay. A description of this type of model is given by Milter (1986).

The case concerns the Eastern European Distribution Organisation (EEDO) of ICL, Britain's largest computer firm. EEDO was created in 1978 to sell ICL products in Greece and six Soviet bloc countries. By 1984 they were operating at a modest profit, but were facing increasing competition from IBM and other European suppliers of information technology. Substantial variations in the political climate from one country to the next meant that they faced considerable risk in some countries, exacerbated by the need for counter-trade agreements since currency could not be exported. Overall, EEDO was experiencing increasing trading difficulties, and the territory manager felt that it was worth reconsidering the strategies being pursued by his country managers.

3.1 The EEDO model

The decision conference was attended by the manager of the EEDO territory, the seven country managers, and a staff person for each country. During the initial problem formulation stage, each country manager stated his current strategy; Greece, for example, was operating as a small local branch. These are subsequently referred to as the 'Status Quo' strategies.

Each country manager was then asked to formulate plausible alternative strategies, some driven by a cut-back in total resource for the country, but others responding to the availability of more resource. It was agreed that the key resource was operating expenditure, broken down into service and country operating expense. Capital expenditure was also considered, but was given no weight when the model was run. An example of these alternative strategies, for Greece, can be seen in Figure 1.

Figure 1: Costs and benefits associated with possible strategies for Greece.

VARIABLE 1: GREECE

	COST			BENEFIT			
	C/OPX	S/OPX	CAPEX	PROFIT	REV	SAFTY	FUTUR
1 CARE AND MAINTENANCE	10	30	0	0	0	100	0
2 STATUS QUO	195	105	15	10	10	75	15
3 LARGER ICL BRANCH	375	120	30	25	30	15	25
4 + LOCAL ASSEMBLY	465	210	45	45	50	45	70
5 + DM1	765	240	145	70	75	20	90
6 + GOVERNMENT SECTOR	1125	270	145	100	100	0	100
WITHIN CRITERION WTS				29	57	40	70
ACROSS CRITERIA WTS	100	100	0	100	50	60	70

The first strategy, a cut-back to care and maintenance, operates at less resource than the status quo, while strategies 3 through 6 identify opportunities that require more resource. If the manager had more resource, he would operate as a larger ICL branch; with even more resource, he would add some local assembly of products; more resource would enable him to sell the DM1 computer, and with even more, he would open up markets in the government sector, for at that time the branch was selling only to the manufacturing and retail trade. Yugoslavia's strategies are shown in Figure 2.

Figure 2: Costs and benefits associated with possible strategies for Yugoslavia.

VARIABLE 2: YUGOSLAVIA

	COST			BENEFIT			
	C/OPX	S/OPX	CAPEX	PROFIT	REV	SAFTY	FUTUR
1 AGENTS ONLY	20	30	0	17	0	100	0
2 + ICL HELP BRANCH	180	320	0	8	3	95	10
3 SQ + ICL DIRECTION	790	600	40	0	5	90	0
4 + LT COOPERATION	1075	650	80	71	78	0	70
5 + JOINT VENTURE	1300	500	250	100	100	85	100
WITHIN CRITERION WTS				100	100	100	100
ACROSS CRITERIA WTS	100	100	0	100	50	60	70

They range from an agent's only operation to a vigorous strategy that involves local long-term cooperation with central government plus a joint venture with some Yugoslavian company. While the model was generated by the group, it was drawn by the facilitator on large white boards that were easily visible by all participants, and it was also implemented on a computer.

3.2 Costs

The costs shown in Figures 1 and 2 were determined by the managers on the spot. Referring to available data, the manager of Greece said his current costs were £64,000 a year in operating expense, and £35,000 in service operating expense, so over a three-year period the figures would be £195,000 and £105,000, respectively. Then the manager of Greece judged what it would cost him to operate as a larger ICL branch, to add local assembly, and so forth.

Thus, aside from the Status Quo strategies, most of the cost numbers are fairly well-based judgements, and the managers assess them in front of their colleagues, providing justifications for the figures. There is a lot of discussion during this period as the managers challenge each other, surfacing assumptions and generating more satisfactory assessments that are acceptable to the group. At this stage a group consensus is beginning to develop, and the rationale is as important as are the numbers themselves.

3.3 Benefits

Participants were next asked about the goals of EEDO. Obviously, they had a profit goal, but they were also interested in the size of revenues, both within the next three years. They also considered the potential for generating profit and revenue in the next three-year period, and included this under the heading 'future potential'. It was important to include future potential as a long-term objective because it is possible over a shorter period to milk a country for profit and revenue, leaving it so weak that it is unable to generate profit and revenue in the long term. Finally, managers were aware that some strategies were riskier than others, so they included 'safety' which is the inverse of risk. A decision analyst would typically incorporate this as probability distributions, but I find that this usually increases the complexity of the problem to an unacceptable level. It is simpler to incorporate risk as a criterion dimension, provided that it is reasonably well defined. Here it was considered as a combination of regret and probability. How seriously would you regret it if you didn't achieve the profit and revenue that you expect to achieve by adopting that strategy? Then, weight that regret by its likelihood of occurring to obtain an indication of preference. Low regret that is unlikely to be experienced is most preferred, while high regret that is likely is least preferred.

3.4 Assessing benefits

Each country manager, assisted by his colleagues, judged the potential benefits of each strategy separately for the four criteria, profit, revenue, future potential and safety. Assessments were made on preference scales, which are relative scales where 0 is associated with the least preferred strategy and 100 corresponds to the most preferred strategy. For Greece, the care and maintenance strategy was judged to yield least profit, with the top strategy, larger branch plus local assembly, DM1 and government sector, providing most profit (see Figure 3).

The status quo was positioned at only 10 relative to the two extreme strategies. Thus, moving from the status quo to the top-level strategy was judged to provide an increase in value associated with profit, over three years, that would be nine times as large as the reduction in value associated with profit by cutting back from the status quo to care and maintenance. The remaining strategies were then located on

Figure 3: The profit preference scale for strategies in Greece.

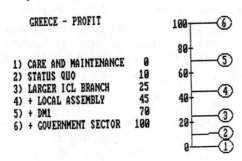

the scale such that the differences in value between strategies were reflected in the differences between the numbers. By comparing ratios of these differences, an interval scale of preference was generated.

The process continued for the remaining three benefits, with no attempt at this stage to compare the separate scales. Note that the safety scale runs in the opposite direction from the other three: care and maintenance is safe, the most costly strategy is the riskiest. Once the assessments were complete for Greece, the group turned to the manager of Yugoslavia, and his cost and benefit scales were judged. In turn, all country managers assessed scales for their strategies.

Considerable discussion accompanied the generation of these scales. Frequently, participants uncovered assumptions that others were making. For example, the profit scale for Yugoslavia reveals that a cut-back can generate some profit in the short term, though that is not good in the long term. When participants disagreed about the benefits, the facilitator stimulated discussion to uncover the reasons for the differences in opinion. In general, only group process techniques that build consensus were used, so when assessments could not be agreed, voting or taking averages were not used. Instead, disagreements were starred on the whiteboard, to be the subject of later sensitivity analyses.

3.5 Assessing weights

The result of repeating this process for all seven countries is to generate 28 (7 countries by 4 benefit criteria) separate non-comparable preference scales. The scales were made comparable by using weighting techniques associated with multi-attribute utility (value) theory

(Edwards and Newman, 1982). The group was asked to compare the difference in the values associated with profit at the least preferred strategy and the most preferred strategy in Greece with the difference between the least and the most preferred strategies in Yugoslavia. Comparisons of all the other five countries identified the biggest difference in profit: it was in Yugoslavia so this scale was assigned a weight of 100.

Next, the group was invited to assign weights to the profit differences in the other countries. It was explained that these weights reflect the actual difference in profit and also how much the group cares about that difference. An equivalent difference might be observed in two countries, but it was possible to care more about the profit difference for one of the countries for various business or other reasons. These weights are shown in Figures 1 and 2 (see Section 3.1) as 'within criterion weights'. For example, the profit difference for Greece is 29% of the profit difference for Yugoslavia.

Finally, a similar comparison process was used to generate weights that compared the benefit criteria, the 'across criteria weights'. These reflected the relative importance of the criteria; the relativity of the numbers was emphasised during the assessment process because people have a tendency to try to judge these in an absolute way, whereas they only reflect the value differences between the most- and least-preferred strategies.

Thus, two weights are associated with each scale; one set, the 'within-criterion weights', makes the 7 scales for a given criterion comparable across the 7 countries; the other, the 'across-criteria weights', makes the 4 scales for each country comparable across criteria.

Various cross-checks were made to ensure the internal consistency of the weights. For example, a 50-point difference on a scale weighted 100 should be equivalent to a 100-point difference on a scale weighted 50. Several changes to the weights were made by the group as a result of asking a variety of internal-consistency questions.

To ensure that the assessed weights were meaningful to the group, a paired-comparison technique, similar to that described by Sayeki (1972), was used. This helped participants to understand: that these 'swing weights' (as they are sometimes called) express the change in value as one 'swings' from the least-preferred to the most-preferred strategy, that only the ratios of weights to one another are meaningful, and that the weights equate the units of measurements from one scale to the next and also express relative importance.

3.6 Results

The computer first normalises the within-criterion weights so that all seven weights sum to 1.0 for a given criterion. Next it normalises the four across-criterion weights. Then it multiplies each benefit scale by the two normalized weights associated with it, and adds the four doubly-weighted scales for each country to give a single benefit scale. It also adds the two cost scales, giving a single cost scale for each country. Thus, for every strategy there is an associated single cost and a single benefit, and these costs and benefits are comparable across countries.

For Greece and Yugoslavia alone there are 6 x 5 = 30 possible combinations of strategies. Including the other five countries brings the total to 105,000. Each one of these possible territory strategies will be referred to as a 'package' of individual country strategies. For each package, a total cost and total benefit can be obtained by adding the individual costs and benefits of the component country strategies. For example, one such package is formed by operating the status quo strategy in each country. The total cost and benefit of this package are shown in Figure 4.

Figure 4: The proposed package: Status Quo in all countries.

VARIABLE	COST	WTS	BEN	LEVEL	
			PROPOSED PACKAGE		
1 GREECE	300	121	10	STATUS QUO	(2 OF 6)
2 YUGOSLAVIA	1390	316	0	SQ + ICL DIRECTION	(3 OF 5)
3 USSR	325	109	21	STATUS QUO	(2 OF 7)
4 POLAND	840	171	15	STATUS QUO	(2 OF 5)
5 CSSR	1190	151	95	STATUS QUO	(2 OF 5)
6 HUNGARY	550	75	23	STATUS QUO	(2 OF 5)
7 BULGARIA	420	57	21	STATUS QUO	(2 OF 4)
	5015		185		

This defines the current operation in EEDO: it costs about £5 million over three years, for a relative benefit of 18.5, where 0 corresponds

to the overall benefit of operating the least-beneficial strategy in each
country, and 100 corresponds to following the most beneficial strategy
in each country.

Figure 5 shows the location of this status quo package relative to the
other 104,999 packages. All the 105,000 packages are located in the
shaded portion; there are no low-cost, high-benefit packages. The
curve shows the locus of the best packages for a given cost; it is an
efficient frontier, and there are only about 20 packages located on it.

Figure 5: The efficient curve and the location of the Proposed package, a Better one
         and a Cheaper one.

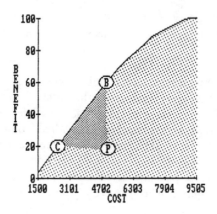

The status quo package is indicated by the point P (Proposed), but
two improved packages are also shown: point B (Better) provides more
benefit at no extra cost, and point C (Cheaper) provides the same
benefit as the status quo but at lesser cost. If package P is
acceptable, then both packages B and C, which are improvements,
should be attractive to the group.

The changes in country strategies that are required if either B or C is
chosen are shown in Table 1.

The 'levels' refer to the strategies for each country, so higher levels
imply greater cost. If movement from P to B is considered, note that
cut-backs are required in three countries, Greece, Russia and Bulgaria
with the freed resource given to the other four countries. The loss of
benefit from the three countries will be more than made up by

pursuing more vigorous strategies in the other five. Of particular concern to the group was the observation that they were wrongly

Table 1: Changes in the country strategies from the P to either the B or C packages.

VARIABLE	1	2	3	LEVEL 4	5	6	7
1 GREECE	CB	P					
2 YUGOSLAVIA	C		P		B		
3 USSR	CB	P					
4 POLAND		P		CB			
5 CSSR	C	P	B				
6 HUNGARY		P	CB				
7 BULGARIA	CB	P					

positioned in every country, that whether the B or the C package was chosen, strategies would have to change everywhere.

This was not the final model; it was an initial result that stimulated considerable discussion, leading to changes in the model. After several sensitivity analyses, in which differences in judgements were explored (and usually found to have no effect on the move from P to B), the group accepted the new model. They then explored the efficient curve in the vicinity of the B point, and realised that with a slight increase in total resource, they would only have to cut back in two countries. The case for more resource was subsequently put to the European Division, of which EEDO was a part, and the additional resource was granted.

New strategies were put in place within a few months of the March 1984 decision conference. By the end of 1985, revenues, orders and profits had all doubled. The improvement in performance was attributed by the EEDO manager to the new strategies formulated at the decision conference. By mid-1986 enough extra resource had been generated that the two countries which suffered cut-backs were re-injected with funds so they could grow.

4. DISCUSSION

I have argued elsewhere (Phillips, 1984) that models like the EEDO model are 'requisite': they are sufficient in form and content to solve the problem at hand. The EEDO model represented the shared social reality that was created by the group. Any one individual would have had a more detailed understanding of his part of the problem, so the model is a simplified representation of each individual's perspective on the problem. Yet by combining many perspectives in one model, new meta-perspectives emerge, like the realisation that each country was wrongly positioned. The model is generative in that an existing situation is described in such a way that new insights emerge, and that it is possible to see how the situation can be transformed into a new social reality (Gergen, 1982). Requisite decision models always involve human activity systems, so while they may be descriptive of a currently-perceived social reality, and always suggest how that reality may be transformed (e.g., the possible move from P to B or C in the EEDO model), they are rarely predictive of what people will actually do. The model is only a guide to action, not a normative prescription, and it is at best conditionally prescriptive. It serves to bring about a shared understanding of a problem, and to create a commitment to action.

In addition, the model differs from purely descriptive models in that it participates in the reality it is supposed to model. As the results of the modelling become available to the participants, their understanding of the problem changes and deepens, and these new perspectives are reflected back as changes to the model. Thus, the model-building process is interactive and iterative, the social reality that is shared by the group changing as the model is revised. Eventually the process stabilises, and participants are both satisfied with the model and unable to derive any further insights from it. At that point, major conclusions and issues are summarised in the decision conference and an action plan is drawn up. The model has served its purpose, and the group returns to its usual ways of communicating without the model.

Applying requisite decision modelling to projects that involve multiple stakeholders requires that the differing perspectives be included in one model so that the effect of differences of opinion on the options being considered can be examined. This is relatively easy to do with models that represent value judgements as preference scales and the relative importance of decision criteria as numerical weights. It is extremely difficult to do with verbal arguments alone.

For example, in the case of EEDO, one of the participants argued that the group was giving too little importance to long-term considerations. With a weight of 100 on profit and 50 on revenue, both over a three-year period, there is a total weight of 150 on the first three years as contrasted to only 70 on future potential, representing the following three years. To test his contention that some of the cut-back countries would be revived if more weight were put on the longer term, the weight on future potential was increased to 150, with all the other weights remaining the same. The result on the P, C and B strategies is shown in Table 2, which should be compared to Table 1 (see Section 3.6).

Table 2: P, B and C strategies if the weight on future potential is increased from 70 to 150.

VARIABLE	1	2	3	LEVEL 4	5	6	7
1 GREECE	CB	P					
2 YUGOSLAVIA	C		P		B		
3 USSR	CB	P					
4 POLAND		P		CB			
5 CSSR	C	P		B			
6 HUNGARY		P	CB				
7 BULGARIA	CB	P					

Notice that cut-backs occur in the same three countries, and all other strategies are unchanged as well, except that a higher level is recommended for the B strategy in Czechoslovakia. Thus, more weight on the long term has virtually no effect on the suggested action plan. It is through requisite decision models like this one that the effect on action of differences of opinion can be examined. Often, actions can be agreed even though participants disagree about various aspects of the problem. Even when there is an effect on actions, the model will suggest how consensus can be reached without compromise, and with little loss of benefit.

Decision conferencing is not the only way to create requisite decision models. Any iterative and consultative approach to model building, such as value-oriented decision analysis (Chen and Mathes, this volume), can be used. The general point of this paper is to suggest that requisite decision models can be useful for large-scale technological projects.

5. CONCLUSION

The model developed in a decision conference lends structure to thinking, and allows all perspectives on a problem to be represented and discussed. This helps to take the heat out of arguments that seem to arise from differences of opinion. Thus, the process facilitates communication among participants, providing "a way to talk differently", as one person put it, and it surfaces assumptions that are often different from one person to the next. Because the model developed by the group shows what the organisation can do, rather than just describing what it does do, creative and lateral thinking is facilitated. Overall, decision conferencing helps to build a sense of common purpose.

More generally, requisite decision models, whose form and content are just sufficient to solve a particular problem, can contribute to the solution of problems that arise from the proposed introduction of large-scale technologies. In particular, controversial decisions involving different groups who do not agree with one another can be informed by sensitivity analyses on requisite models that combine the differing perspectives of the interest groups.

Effective decision making requires a balance between content, structure and process. Content refers to data, information and value judgements that are relevant to the problem at hand; structure shows how the items of content are related; and process concerns how content and structure are generated and used in taking decisions. If any of these three elements is neglected, the quality of decision making suffers. Requisite decision modelling is an approach that helps to keep all three in balance.

Organisations using decision conferencing report that the process helps them to arrive at better and more acceptable solutions than they can achieve using their usual procedures, and agreement is reached much more quickly. Many decision conferences have broken through stalemates created previously by lack of consensus, by the complexity of the problem, by vagueness and conflict of goals, and by failure to think creatively and freshly about the problem.

Experience suggests that decision conferencing works best in organisations when four conditions are met reasonably well. First, consultation should typically precede decision making. The participative style of decision conferencing is sometimes resisted by authoritarian leaders. Second, communication links should exist across the organisations' divisions and sub-divisions, so that information flows laterally as well as vertically. This helps to ensure that the information needed in a decision conference will be available. Third, a climate of problem solving should exist in the organisation, so that options can be freely explored. If the organisation's style encourages the manipulation of options to serve pre-determined solutions, then decision conferencing will be resisted because new, un-anticipated solutions may emerge. Finally, authority and accountability should be well-distributed throughout the organisation, neither concentrated at the top nor totally distributed toward the bottom. This ensures that participants in a decision conference hold the authority to carry out the action plan that they create.

Since many large-scale technological projects involve public-sector organisations, the above conditions raise questions about the suitability of decision conferencing for government institutions. For example, civil servants may find it impossible to persuade ministers to participate in a decision conference; representatives of interest groups may not wish to meet face-to-face; leaders of public opposition groups may not be willing to participate. However, if one sees the task of dealing with large-scale technological projects as requiring strategic planning, whose purpose has been described by Bryson and Einsweiler (1987) as "...to help key decision makers figure out what the role of government ought to be, what it should do, and how it should allocate limited resources", then the relevance of the EEDO case to government planning should be evident. Whatever approach is used, many stages in the public-sector strategic planning process can be aided by decision conferencing. These include clarification of the organisation's vision, mission or goals; analysis of strengths and weaknesses, opportunities and threats; identification of strategic issues, formulation of possible strategies, and evaluation of options. By creating a requisite model of the particular problem at hand, civil servants can role-play the different perspectives they think would be adopted by problem-owners or stakeholders who can not be accessed directly. The development of strategies that are robust with respect to different stances can be facilitated by this process. In particular, if this is done in the early stages of the project, it may be possible to anticipate subsequent difficulties, and help politicians adopt a flexible approach.

REFERENCES

Bezembinder, Th. (this volume). Social choice theory and practice.

Bryson, J.M. and Einsweiler, R.C. (1987). Strategic planning: Introduction. *Journal of the American Planning Association, 53,* 6-8.

Chen, K. and Mathes, J.C. (this volume). Value oriented social decision analysis: a communication tool for public decision making on technological projects.

De Hoog, R., Breuker, E. and Van Dijk, T. (this volume). Computer assisted group decision making.

Edwards, W. and Newman, J.R. (1982). *Multi-attribute evaluation.* Beverly Hills and London: Sage Publications.

French, S. (1987). *Decision theory: An introduction to the mathematics of rationality.* Chichester: Ellis Horwood Ltd.

Gergen, K.J. (1982). *Toward transformation in social knowledge.* New York: Springer-Verlag.

Howard, R. (1975). Social decision analysis. *Proceedings of the IEEE, 63(3),* (pp. 359-371).

Low, K.B. and Bridger, H. (1979). In: B. Babington-Smith and B.A. Farrell (Eds.), *Learning in small groups: A study of five methods.* London: Pergamon Press.

Milter, R.G. (1986). Resource allocation models and the budgeting process. In: J. Rohrbaugh and A.T. McCartt (Eds.), *Applying decision support systems in higher education.* San Francisco: Jossey-Bass.

Phillips, L.D. (1984). A theory of requisite decision models. *Acta Psychologica, 56,* 29-48.

Sayeki, Y. (1972). Allocation of importance: an axiom system. *Journal of Mathematical Psychology, 9,* 55-65.

Vári, A. (this volume). Approaches towards conflict resolution in decision processes.

Wilke, H.A.M. (this volume). Promoting personal decisions supporting the achievement of risky public goods.

VALUE ORIENTED SOCIAL DECISION ANALYSIS: A COMMUNICATION TOOL FOR PUBLIC DECISION MAKING ON TECHNOLOGICAL PROJECTS[1]

KAN CHEN AND J.C. MATHES

College of Engineering
University of Michigan, Ann Arbor

1. INTRODUCTION

Many public policy problems at all levels - national, regional, state, and local - involve issues that can not easily be reduced to simple either-or propositions. That is, the problem involves diverse constituencies and interest groups who can not be grouped into two sides so that trade-offs can be made and a resolution reached through the typical political process (in the United States). For example, at the local level such an issue would involve extending the runways of the local airport, building a shopping center in a suburban area, or combining the fire and police departments into one organization. In such situations, the usual political alignments do not apply, as Republican and Democratic council-persons will align themselves on both sides of the issue because typical party loyalties among the citizens do not yield predictable stands on the issue.

The tendency in such situations seems to be to oversimplify the issue so that it fits into the typical political decision making process. Various means of oversimplification result. One is to simplify the issue in such a way that it becomes one of the interests of the community as a whole versus the interests of the neighborhood. An other is to define it in such a way that the various interest groups logically fall into two opposing coalitions so that some sort of negotiation or trade-off process - or even a simple play of power - can ensue. A third is to turn the issue into an ideological issue - one which conforms to the perceived ideologies of the Republican and Democratic parties, one which conforms to an accepted opposition of values in the community (such as pro-business or development versus status quo or quality of life), or one which conforms to a pervasive ideological conflict that manifests itself whereever possible on any specific issue (such as the pro- and anti-nuclear energy debate [DelSesto, 1980; Nelkin and Pollak,

111

Ch. Vlek and G. Cvetkovich (eds.), Social Decision Methodology for Technological Projects, 111–132.
© *1989 by Kluwer Academic Publishers.*

1980; Vlek, 1986]).

The result is a definition of a problem in terms that are amenable to political resolution. Oversimplification, however, can result in a solution that does not reflect the values of as many of the citizens that it possibly could have or that might not be the most appropriate for the problem. Even without oversimplification, an adversarial situation is often established that can lead to less than desirable results.

What is needed is a means of defining the problem and the related issues in such a way that reflects its complexity as well as the various values of the multiple interest groups involved. Such a means would recognize that various persons and groups have different understandings of the problem, different interpretations of the possible solutions, and different values for evaluating those solutions. This being done, then these diverse interests need to be represented or involved in some stage of the policy making process so that the resolution of the issue is appropriate to the problem and represents a multiplicity of social values.

We have developed and applied a Value Oriented Social Decision Analysis (VOSDA) methodology that addresses these needs. The purpose of this method is to improve the effectiveness of decision making on public policy issues by enhancing the communication processes involved. The goal is to help formulate a policy or decision on a public issue where the nature of the problem and the formulation of the issues involved are somewhat unclear. The method aims to foster mutual understanding of issues and alternatives. Its application both structures complex problems and encourages multiple parties at conflict to develop and share quantitative information about their perceptions and trade-offs at successive stages of the public decision making process.

The method draws upon multiple-criteria decision making techniques to increase communication among multiple interest groups and multiple decision makers in a public decision making process. To that end we have added a significant communication component to a traditional decision analysis procedure. Decision analysis involves identifying alternative actions and possible results, judging the likelihoods of those results occurring, establishing the desirability of the possible results, and determining the preferred policy. We expand this procedure to include the relevant public agencies, public groups, and private parties involved in an issue; then we establish on-going communication among them throughout the procedure. The result is to use a Value Oriented Social Decision Analysis (VOSDA) methodology primarily as a communication tool for public policy making.

2. GOALS AND ASSUMPTIONS

The VOSDA methodology assumes a fundamentally rational approach to public policy making, both descriptively and normatively, but one with a process rather than output focus. It combines the rational social decision analysis methodology (Howard, 1975) with a process emphasis that assumes the viability of interpersonal group dynamics procedures (Hammond and Adelman, 1976). It furthermore focusses on the formal communication process involved, drawing from management communication (Mathes and Stevenson, 1976) and tagmemic rhetoric (Young, Becker and Pike, 1970, which was influenced by the communication approaches of Carl R. Rogers and Anatol Rapoport). It includes a participatory component that operationalizes the input of the public(s) much more explicitly than does the public hearing model (and also perhaps more constructively than the public hearing model used in nuclear power plant decision making; cf. Ebbin and Kasper, 1974).

We assume that communication to increase the rationality of the public decision making process while at the same time allowing multiple decision makers and interest groups to interact within the accepted political process will reduce conflict and facilitate a decision that is more widely accepted as well as more optimal in the sense of Pareto optimality than otherwise would be the case (Raiffa, 1982). This mode of communication on a single issue is not necessarily inconsistent with a process such as "vote trading" among multiple issues.

We furthermore have assumed a political context that is familiar in the United States (but which may not be representative of other countries). We assume an interest group or group theory model of the political process (Greenstone, 1975; Salisbury, 1975) which models our two political parties as coalitions of various interest groups rather than primarily as opposing ideological positions. At an operational level, we assume that agreement and harmony among these groups in regards to any policy decision is not requisite or even possible (Irland,

1975). Thus, policy making usually strives for a decision that to varying extents satisfies as many groups as possible (Simon, 1976) through a decision that embodies a diversity of values and/or a variety of trade-offs.

Given this nature of the political process, then, the goal of the VOSDA methodology is to facilitate mutual understanding on public issues among various interest groups, institutions, and policy makers. It aims to create a detailed understanding of a complex issue by the parties that might not otherwise be possible, and indeed goes so far as to make explicit the differences among various interest groups - differences in perception of the problem, recognition of alternatives, and selection and weighting of criteria for evaluation of alternatives. These communication functions can enhance the possibility of mutually acceptable trade-offs in the subsequent public decision making act (Bauer and Wegener, 1977). In general, the goals of VOSDA seem remarkably similar to those outlined by Brehmer (this volume), although VOSDA does not rely on a computerized communication aid.

Thus, as we envision it, VOSDA complements the policy making process in the United States. Its ultimate goal is to improve the rationality and efficiency of that process rather than to modify it in some basic manner. Each of our case study situations was designed on this premise.

3. THE VOSDA METHODOLOGY

The VOSDA procedure assumes that a decision analysis team is in the role of consultant to a policy making body. This team performs the social decision analysis for the body. The procedure was developed with two case study situations. In the first case our project team acted in the role of consultant to the City Administrator of Ann Arbor, who considered to install a solid waste shredding facility to serve the local community. In the second case our team acted in the role of consultant to a Special Joint Committee of the State of Michigan Legislature, which considered alternative strategies for the state's electrical energy future. We present a detailed explanation of the method we followed so that others can evaluate our theoretical assumptions and determine if it can be adapted to other needs or situations.

The VOSDA procedure consists of three stages (cf. Figure 1):

(1) clarification of the problem;
(2) identification of alternative actions and possible effects or attributes; and
(3) determination of the preferred policies.

These three stages involve a repetitive pattern of discussion, summary, revision, and communication. Each stage is marked by communication - the circulation of each person's or group's descriptions or judgments for that stage. The final output of the VOSDA procedure is fed into the actual policy making process, whatever that is for the particular situation. (Our two case study situations modelled different public policy making situations, as described below.)

Figure 1: The three stages of the VOSDA procedure.

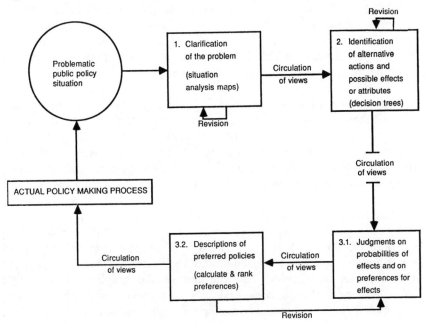

The procedure is initiated with individual discussions by a member of the team with each of the parties with an interest in the issue. These parties are represented by the known and anticipated persons and interest groups with a stake in the issue, including all of the individual politicians (in the case of the city situation, the mayor, city council-persons, selected county commissioners, and selected township supervisors). The result of these interviews is a set of individual statements of their perceptions of the problem. When each individual statement represents to the satisfaction of the person his or her initial understanding of the problem, these statements and their corresponding situation analysis maps, similar to cognitive maps (Axelrod, 1976), are circulated among all persons involved.

The second and third stages consists of a traditional decision analysis procedure, but with the added component of circulation of the views of all persons at the conclusion of each stage. We ask everyone to identify the alternative actions and possible effects ("attributes of consequences" in decision analysis terms), and we circulate those views. Then, we ask everyone to judge the likelihood of those effects (that is, to assign probabilities to them), and to evaluate the desirability of those effects (that is, to assign preferences or utilities to them). With this information, we calculate to their satisfaction everyone's preferred policy. We circulate those statements as well as a summary statement which ranks the various alternative policies according to the preferences of everyone involved.

Throughout the process, everyone has the opportunity to revise and approve their statements and ideas at each stage before these are circulated (as well as to "pass" on any occasion). Invariably, our discussions are with persons individually. The interviews at each stage take from one-half hour to one hour with each person, and sometimes longer if the persons become interested in the technique itself.

The communication or output of each stage is put in graphic form as much as possible, although these techniques still need much improvement. Some of the graphics are illustrated below. They range from standard decision trees, enhanced to illustrate a personal view, to various graphs and tables. The purpose of the graphics is to enable the various views to be compared and contrasted easily. The graphics are supplemented by verbal descriptions and analyses.

4. SOCIAL DECISION ANALYSIS (VOSDA) PROCEDURE

The following outline illustrates our actual application of the method (detailed explanations for each case situation are in Chen, Mathes, Jarboe and Wolfe, 1979, and Chen and Mathes, 1986). Figures illustrate the outputs of several of the stages, but because of length constraints we do not include them all.

4.1 Clarification of the problem

4.1.1 The VOSDA team discusses the problem with each person or group individually.

4.1.2 The VOSDA team prepares a descriptive summary of each person's or group's analysis of the problem, and feeds it back only to that person or group for possible revision.

4.1.3 Each person or group revises and edits this summary of his or her description of the problem however (s)he desires.

4.1.4 All of these summaries of the problem, as well as a composite summary, are circulated among everyone else before the next stage of the process is initiated.

The summaries of the problem are in the form of "situation analysis maps": Figure 2 presents the composite map for the second case study. Each individual perception of the problem is highlighted separately on a copy of the composite map. Both the composite map and the set of individual situation analysis maps are circulated.

Figure 2: Situation analysis map for Michigan's electrical energy future alternatives.

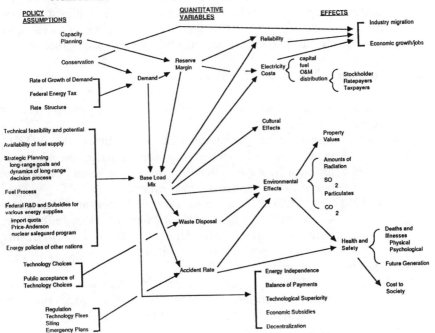

4.2 Identification of alternative actions and possible effects or attributes

4.2.1   The VOSDA team outlines a skeletal decision tree, using the information obtained during the interviews and presented on the situation analysis maps, supplemented by any other relevant information (such as testimony, report, or presentation).

4.2.2   The VOSDA team asks each person or group to identify: (1) the various alternative actions possible that can be taken to address the problem, and (2) the possible effects (attributes) of those actions.

4.2.3   The VOSDA team constructs individual decision trees and a composite decision tree, which includes all of the attributes by which to evaluate a possible action (policy).

4.2.4   Each person or group revises their individual decision tree (based on the composite tree) to their satisfaction.

4.2.5   All of the individual decision trees as well as the composite decision tree are circulated.

Figure 3 presents a specimen individual decision tree from the solid waste shredding facility case. As with the situation analysis maps, the individual decision tree is outlined on the composite decision tree. However, to simplify the process each individual's attributes are not identified until stage three.

4.3 Determination of preferred policies

4.3.1   Judgments on probabilities of effects and on preferences or utilities for effects.

    4.3.1.1   Based on the composite list of attributes, the VOSDA team asks each person or group to judge the likelihood of each possible effect occurring, as some effects are more likely to occur than others.

    4.3.1.2   The VOSDA team then asks each person or group to select and weight the desirability of those effects (or attributes) they consider important for evaluating alternative policies on this issue.

4.3.2   Description of preferred policies.

    4.3.2.1   The VOSDA team performs a decision analysis for each individual person or group to calculate and rank their preferences for the policy alternatives identified on the composite decision tree.

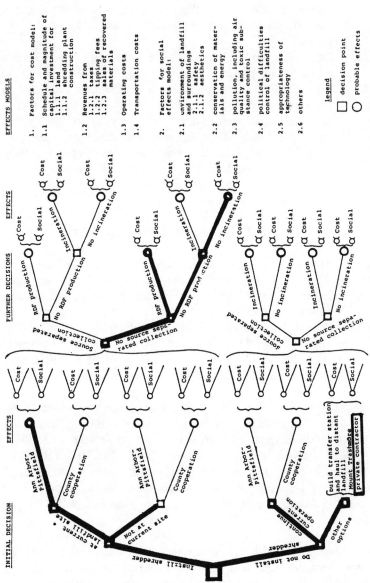

Figure 3: Specimen individual decision tree (highlighted on the composite decision tree).

Note: Purchase of sufficient land for operation of each option is assumed in the cost and social effects models.

*Specific design decisions such as creation of a buffer zone around the landfill, extension of current barriers, location of landfill entrances, condition of access road and siting of facility within landfill will also be explored in the cost and social effects models.

4.3.2.2   The VOSDA team reviews with each person or group the results of their individual decision analysis. The results are revised (which requires a revision of the results of the prior stages for that individual) until that person or group feels comfortable with the results.

4.3.2.3   The collective results of the decision analyses are circulated in graphic and verbal form.

4.3.2.3   The VOSDA team prepares a descriptive analysis, as appropriate, of the similarities and differences among the individual/group decision analysis results. This descriptive analysis does *not* include some sort of group decision rule to recommend an optimal policy. This analysis as well as the collective results of the individual/group decision analyses provide input into the actual policy making process.

5. PROCEDURAL APPLICATIONS AND VARIATIONS

The actual VOSDA procedure was much more involved than this brief outline suggests. Many of the steps in the process evolved only after considerable deliberation and, in some instances, investigation and evaluation of alternative possibilities. Most of our concerns and deliberations have not been discussed in the publications cited above, but some are relevant to this discussion. A few seem particularly relevant to a comparison of the assumptions and goals of VOSDA with other social decision methodologies for technological projects.

From the outset, we considered the first stage of the procedure to be the most important stage methodologically, other than the communication/group interaction component of the entire procedure. Typical decision analysis, and indeed the typical decision making or policy making procedure, accepts the problem or issue at hand as a given. Theoretically, however, this is not the case, for a problem is a subjective reality (Lindblom, 1968), or at most a shared social reality (Phillips, 1984).

The way a problem is defined predicates the alternatives and the solutions. We therefore felt that to model the ideal problem solving process, the VOSDA procedure should begin with an exploration and definition of the problem by all persons or groups concerned with a problematic situation. The second case study verified this assumption.

The issue initially was framed by the legislature as an issue of nuclear safety, but by the time we entered the scene it had become one of alternative electrical energy futures. Nuclear safety became one of the many attributes by which to evaluate an energy future scenario; it would have been so even if the energy future scenario had been specifically designed around safety issues.

An additional benefit of introducing a problem definition stage is the participation of the interest groups in the initial definition of the issue to be addressed. Although VOSDA does not provide a mechanism for them to participate in final policy making, it does involve them in a participatory manner in this important formative stage of policy making. Substantive involvement early on theoretically could be more constructive and therefore satisfying than being asked to vote "yes" or "no" on an issue already framed by the policy makers.

Perhaps because of the communication component of our procedure, considerable effort was expended on developing a means of defining the problem that would synthesize multiple perspectives in graphic as well as verbal form. Originally, we had planned on using decision trees to structure the participants' perceptions of the problem. We soon realized, however, that a decision tree is not adequate to capture an individual's or group's perception of reality; it is formally structured in either-or, sequential choices that logically beg definition of the problem. As Waller (1976, p. 1) said, "in real-world assaults on complex problems, one begins with 'unstructured states of confusion', to use a phrase of William James', not with decision trees or numerical algorithms". Drawing on previous experience with simple qualitative cause-effect or relational models (Chen and Lagler, 1974) as well as on environmental models (Holling, 1978; Johnson, 1980), therefore, we decided to adapt the concept of the cognitive map (Axelrod, 1976). A cognitive map is a pictorial representation of the causal assertions of a person as a graph of points, which represent the concepts a person uses, with connecting arrows, which represents his or her perceived causal links between these concepts. We labelled our variation as a "situation analysis map", that is, a conceptual arrow diagram of a problematic situation describing the perceived causal relationships among a number of events and anticipated effects. At this stage the relationships suggested are those of a signed digraph, that is, directions (positive or negative) of the effect.

The situation analysis map, therefore, clearly is qualititative and therefore, strictly speaking, can not be labelled a "cognitive map". It does not attempt to quantify the causal relationships at this stage. It only suggests possible correlations and directions between observed (or interpreted) events and anticipated consequences, with intermediate chains. For the purpose of VOSDA, we think this is quite sufficient. It essentially is a heuristic by which to gather concepts and information to construct a decision tree and identify relevant attributes. Quantitative data gathering and analysis follows in the decision analysis procedure; in these subsequent stages many of the relationships can be transformed into weighted digraphs. The situation map serves to describe the complexity of a problematic situation rather than to reduce it to a logical or syllogistic form. It broadens rather than circumscribes perspectives.

The decision tree for the problem and the subsequent stages of the decision analysis then is constructed by the VOSDA team. The tree is designed on the basis of the composite situation analysis map as well as interviews with the participants to identify all of the possible alternative actions and the possible results of those actions. Construction of the tree clearly is a major intervention by the team, as it structures the perceptions of the participants for them. The interview technique, which required a visit prior to the design of the tree and a visit afterwards to check and refine it so that it satisfied each participant, is an essential element in the design of the tree. In each of our case study situations, all of the participants were satisfied that the decision tree designed by the VOSDA team satisfactorily captured all of the major decision alternatives in an acceptable sequence.

To perform the decision analysis for each participant, we used the additive multi-attribute utility model (Keeney and Raiffa, 1976):

$$U = \sum_i \sum_j w_i p_{ij} u_{ij}, \text{ where}$$

U is the program's overall expected utility
w_i is the weight assigned to attribute i
p_{ij} is the probability of event j in attribute i
u_{ij} is the utility of event j in attribute i

The overall expected utility for an option consists of the sum of the expected utilities of that option for each attribute, adjusted according to the weights assigned to the attributes. The expected utility of an option on a single attribute is the sum across all uncertain outcomes of the probabilities times the corresponding utilities. It is important for the participants to realize that their preferences for various options are based on a sum of their preferences for those options attribute by attribute (i.e., they are not directly voting for options themselves as single entities).

The decision analysis for our second case study illustrates the importance of evaluating options in terms of individual attributes. The composite rankings revealed that the participants tended to cluster in two groups (see Figure 5, Section 4.3), which could indicate a polarity in keeping with an ideological or two-side model of the political situation. However, an analysis of the relative importance of the individual attributes in considerations of alternative energy programs indicated that each participant had a distinctively individual position. These individual positions did not conform to the two composite clusters (see Figure 4, Section 4.3); each individual chose an unique cluster of attributes to weight, and each made different trade-offs among the attributes. This result, therefore, seems in keeping with an interest group model rather than a polarity model of the political process.

We treat the probabilities and utilities (and preferences) of each attribute as independent of the others in order to simplify the process. Rather than determining all possible outcomes or combinations, the certainty equivalent of each independent attribute is calculated and then combined according to the formula. If the probabilities and utilities of the attributes are considered dependent, the analysis would be too complicated to serve as a communication tool (and perhaps even for an effective decision analysis).

In keeping with our assumption that the role of VOSDA is to complement the traditional political process, the output of each case study consisted of descriptive (and graphic) representations of each participant's position on the policy alternatives as determined by the decision analysis. These positions, of course, are the participants' "public" positions, rather than their private ones that they might not be willing to express in a public forum such as a hearing, a letter to an elected representative, or a VOSDA-type analysis. We assume that their public statements operationalize their positions, that is, provide the input which influences the policy making process. The diversity, rather than polarity, of opinion then is fed into the normal political policy making process. This diversity is appropriate because it more nearly reflects the complexity of the actual situation.

Presumably, this diversity of public positions would enable various subsequent trade-offs among the participants to be considered that might yield a Pareto superior decision (see Bezembinder, this volume). At the least the process provides for all interested parties to contribute substantively to the analysis and evaluation of the policy alternatives, although they do so in a rational, public mode. In our two case study situations, the VOSDA procedure was completed with a public report that presented all participants' individual positions. We have yet to apply it in a situation where we can observe and analyze its direct impact on the policy making process.

The communication models assumed in the case studies differed considerably, and need to be clarified in further development of the VOSDA procedure. For the city waste disposal situation, the model was one of interpersonal, intragroup communications. The involved parties included all of the persons and interest groups, including the decision makers (e.g., mayor and ten council-persons), with any interest in the issue. Although the final decision was to be made by the decision makers, in the VOSDA procedure all of the interested parties interacted with them.

For the state electrical energy future situation, the model was an information input model where the decision makers control the official information flows and the decision making procedure. The decision makers did not themselves participate in the VOSDA procedure. The legislative subcommittee was in a discovery mode, and was soliciting all available information and opinion regarding Michigan's electrical energy future alternatives. The VOSDA procedure structured this input for them. The participants in the VOSDA procedure were chosen as representatives of the various policy positions (five) that could be identified upon analysis of the initial information and composite situation analysis map. Their collective decision analysis results therefore provided only a synthesis of the numerous individual and interest group attitudes on the issue.

The eventual results in each case study situation were inconclusive. In the case of the solid waste shredding facility, the city council voted to put the issue as a bonding proposition for the voters. The voters voted a million-dollar bond issue to build the shredder. Perhaps the VOSDA procedure contributed to this eventual outcome. Six months later, however, the council decided not to proceed with the shredder because of disputes over technical and economic feasibility that resurfaced. A rerun of VOSDA at this time would have been appropriate. The fact that the decision was reversed at a later date reflects the dynamics of the on-going political process, but does not necessarily call the validity of the role of VOSDA in the original decision into question.

On the electric power issue, as so often happens, the legislative committee did not proceed to an actual policy making stage after its hearings and deliberations over Michigan's electrical energy future. The issue was transferred to an other committee, which could eventually issue a bill that will significantly modify the energy policy of the State. The staff assistant of the Special Joint Committee, for whom the VOSDA team acted in the role of consultant, observed that this other legislative committee probably is not interested in any substantive information on the energy issue. Instead, its role probably will be "to force representatives of interest groups perceived ... to be politically significant, to sit down in the same (usually closed) room to see what compromises they can hammer out". He said, of our VOSDA procedure, that "it seems to me that your report, in baring some of the underlying bases for various positions held by some significant interest groups, could be useful in this fashioning of compromises" (W.J. Scanlon, personal communication, April 27, 1981). For the present, however, the VOSDA report is in the file along with all of the other information and reports collected by the Special Joint Committee.

6. EXPERIENCE AND EVALUATION

Partly because of the long process in applying VOSDA and partly because of the lack of resources, no other full-scale application of the VOSDA approach has been tried. Our own experience in these two applications, reinforced by the evaluation of two outside consultants, suggests the following factors that contributed to the limited success of VOSDA and can be somewhat generalized.

First, the VOSDA analysts were able to establish a trust with all interested parties at an early stage. A number of factors probably led to their positive attitude: prestige of the University of Michigan with which all the analysts were affiliated, prestige of the National Science Foundation which sponsored the research, and the feeling of curiosity and novelty on the part of the participants. Perhaps another major factor, which is of more general value, was the non-threatening procedure used by VOSDA. There was no confrontation in a group setting; there was no intention to coerce any party to change its position or to drive the whole group to consensus or to vote; and better communications seemed to be a modest goal that all parties could support, and actually desired.

As to implementation of the approach in the two cases, a great deal of time was spent in tutorial sessions to explain the whole approach and to explain the specific questions about trade-offs, probability assessment, and decision tree structure based on the situation analysis maps. Once these were understood, however, there seemed to be little difficulty for the analysts to get the numbers needed for the analysis. Thus the VOSDA process proceeded to completion in that the communication process was completed and utilized, that is, new information was produced and incorporated in the actual policy making process that it complemented.

The VOSDA procedure as developed so far has its limitations, as our external reviewers pointed out. At present, it is restricted to an explication in public of the positions of diverse interest groups and interested parties on a complex public decision making situation. It clarifies diversity of opinion but has not demonstrably contributed to consensus or resolution. As such, it provides a basis - a "script", as it were - for later bargaining, negotiation, and compromise. Although a VOSDA approach could be used in this subsequent bargaining process, so far it has not been extended to that stage.

At present, significant limitations on VOSDA as a practical process are the potential time and cost involved. We did not estimate the cost of performing VOSDA in each case study situation, but it was considerable and probably would have been impractical if we actually had been paid consultants. The time involved also was considerable. In both situations, the VOSDA team had difficulty in performing the procedure in a timely fashion - that is, so that it actually complemented the real time political deliberation process. Of the two, we think the potential cost is the more serious problem. The time problem resulted because we were developing and refining the methodology as we applied it and because we had up to 80% of our time devoted to other professional activities. In subsequent applications, time might not be a problem.

Although the VOSDA approach has not been applied again in full scale to an other public issue, it has been taught in a number of courses and found its way into student exercises. In addition, the VOSDA approach has been adapted and combined with other approaches to problems akin to social decision making. For example, in a negotiation research project, the VOSDA approach was combined with the principled negotiation process (Fisher and Ury, 1981) and Paretian analysis (Raiffa, 1982) to conduct a retrospective analysis of a past negotiation that had failed. The research project combined the three approaches and provided the two negotiating parties with an improved insight into the conflict situation, and the subsequent negotiation between the same parties has led to an agreement.

In sum, what has made VOSDA acceptable in the applications so far was neither just the decision analysis logic alone nor the social climate alone, but the synergism between the two. First of all, the underlying rationale of VOSDA pressed the interested parties to express their value and reality judgments explicitly without ambiguous rhetoric. In addition, this was done in a non-threatening social climate that conduced the parties to participate in multilogs that enhanced meaningful communications.

7. THEORETICAL RATIONALE AND RELATED APPROACHES

The theoretical motivation in the original development of the VOSDA approach (Chen et al., 1979) was based on consideration of the two major characteristics that distinguish social decisions from individual decisions:

    (1) the presence of multiple decision makers, and
    (2) the nature of the political decision making process.

The first characteristic rules out the possibility of an optimal social decision (Arrow, 1951; Bezembinder, this volume). It also raises a flag to any approach that would require interpersonal comparisons of utility. Theoretically, social decision analysis can be based on N-person game theory (Rapoport, 1970). Practically, however, game theory does not provide useful prescriptions to help in N-person nonzero-sum game situations, which are typical of public policy issues.

A practical retreat from abstract game theory has been to adapt decision analysis (Raiffa, 1968), originally developed for single decision makers to analyze decisions under uncertainty, to social decisions. A straightforward adaptation is to assume that there is an autocratic decision maker in the public sector who can benefit from decision analysis that takes into account the multiple attributes of the consequences of a social decision by converting them to monetary values (Howard, 1975). Alternatively, these attributes need not be expressed in monetary units if multi-attribute utility theory is applied (Keeney and Raiffa, 1976). In either case, however, the problem of different value trade-offs among conflicting objectives by different interested parties concerned about a policy issue, is not addressed.

Other adaptations of decision analysis do acknowledge and stress the differences in value trade-offs among different interested parties, but do not try to capture the differences among interested parties in their reality judgments in terms of probability assessments or in their perceptions of the problem in terms of the decision tree structure (Hammond and Adelman, 1976; Edwards, 1979). The VOSDA approach was developed with the belief that the differences among the interested parties in problem perception, in reality judgment, as well as in value trade-offs, should be captured in a social decision analysis. From this perspective, VOSDA is quite similar to the GroupMIDAS approach discussed by De Hoog, Breuker and Van Dijk (this volume), even though VOSDA does not rely on computer assisted analysis.

The second characteristic of social decisions, the political decision making process, raises the issue of the relative role of any rational analysis vis-à-vis the actual process. The original development of the VOSDA approach was based on a rejection of the extremities of process-dominant and rational-dominant positions. The intent was to combine process and analysis by using decision analysis to help the process of communication among interested parties, with the assumption that improved communication would lead to speedier and more mutually satisfying social decisions.

An interesting framework has been developed recently (Quinn, Rohrbaugh and McGrath, 1985) that clarifies the various ways of melding the process-oriented organizational development perspective and the analysis-oriented engineering/management-science perspective to decision making. Within this framework, the decision conferencing approach (Adelman, 1982; Phillips, 1984, see also this volume) and the VOSDA approach are remarkable in their similarities in trying to unify the two perspectives. Both approaches emphasize participation of multiple parties that influence social decisions and try to "model" problems and solutions on the basis of decision analysis. The situation analysis map and decision tree models of the VOSDA methodology, especially, together seem very similar to what Phillips calls a "requisite decision model", as these present an interpersonal social reality that provides a guide to action (subsequent decision analysis).

However, there are also significant differences between VOSDA and decision conferencing. In terms of implementation, decision conferencing uses real-time computer support for decision analysis in a group setting. By contrast, VOSDA analysts work with interested parties separately on a protracted time scale, and use graphics but not computer support frequently. The fact that we work with the groups separately rather than in concert also renders such methods as brainstorming and nominal group technique (Huber, 1980) for the most part inapplicable to VOSDA. In terms of group dynamics, furthermore, decision conferencing often is emotionally demanding, resulting in consensual decisions as participants change positions and create new solutions; VOSDA participants work with the analysts in a more relaxed mode over real time, resulting in better understandings of the value and reality judgments underlying the relatively unchanged positions taken by all parties.

In actual applications, furthermore, decision conferencing has been applied mainly to multiple parties within a single organization with a common goal, while VOSDA has been applied exclusively so far to interested parties in a public situation that lacks this framework of common values and goals provided by an organization. Decision conferencing usually is used to address problems requiring immediate action, as it shortens the time frame required for problem solving. VOSDA, as we conceived it, is not amenable to such crisis- or urgency-driven problem solving situations.

The basic premises and the complementary characteristics of the two approaches are such that potentially the two approaches may be applied in concert. For example, VOSDA may be applied first to enhance mutual understanding of interested parties, which then can be brought together in a decision conferencing setting to facilitate social decision making. The same decision analysis model as developed in VOSDA may be used to start off the decision conference.

8. FUTURE DEVELOPMENT

When we initially developed our VOSDA proposal, we established the following criteria, in an ascending order of achievement, to help us evaluate our progress:

1. Each interest group sees its own rationale more clearly.
2. Each interest group becomes more capable in conveying and articulating the rationale of its position to the other groups and to the decision makers.
3. Each interest group understands how its position differs from the others' in terms of perception of the problem, choice of alternatives and attributes, probabilities of outcomes, and trade-offs between attributes.
4. The interest groups are able to develop Pareto superior alternatives to the current policy options.
5. The interest groups are able to agree on what further pilot projects to conduct in order to settle some portion of the issues.
6. Some or all of the interest groups modify some of the perceptions and value commitments in their original positions on the technology policy issue.

Our experience, as reported in this chapter and supported by external review, indicates that so far the procedure meets the first three criteria. Negotiation research combining VOSDA with other approaches in one situation has met the fourth criterion. Further application and development are needed to demonstrate the potential VOSDA has to meet the final two criteria.

REFERENCES

Adelman, L. (1982). *Real-time computer support for decision analysis in a group setting: another class of decision support systems.* McLean, VA: Decisions and Designs, Inc.

Arrow, K.J. (1951). *Social choice and individual Values.* New York: Wiley.

Axelrod, R., (Ed.). (1976). *Structure of decision.* Princeton: Princeton University Press.

Bauer, V. and Wegener, M. (1977). A community information feedback system with multiattribute utilities. In: D.E. Bell, R.L. Keeney and H. Raiffa (Eds.), *Conflicting objectives in decisions.* New York: Wiley.

Bezembinder, Th. (this volume). Social choice theory and practice.

Brehmer, B. (this volume). Cognitive dimensions of conflict over new technology.

Chen, K. and Lagler, K.F. (et al.). (1974). *Growth policy: population, environment, and beyond*. Ann Arbor: University of Michigan Press.

Chen, K. and Mathes, J.C. (1986). Clarifying complex public policy issues: a social decision analysis contribution. In: M.J. Dluhy and K. Chen (Eds.), *Interdisciplinary planning: a perspective for the future* (pp. 83-104). New Jersey: Rutgers University.

Chen, K., Mathes, J.C., Jarboe, K. and Wolfe, J. (1979). Value oriented social decision analysis: enhancing mutual understanding to resolve public policy issues. *IEEE Transactions on Systems, Man and Cybernetics, 9*, 567-580.

De Hoog, R., Breuker, E. and Van Dijk, T. (this volume). Computer assisted group decision making.

DelSesto, S.L. (1980). Conflicting ideologies of nuclear power. *Public Policy, 28*, 39-70.

Ebbin, S. and Kasper, R. (1974). *Citizen groups and the nuclear power controversy: uses of scientific and technological information*. Cambridge (Mass.): MIT Press.

Edwards, W. (1979). *Multiattribute utility measurement in a highly political context: evaluating desegregation plans in Los Angeles*. Paper presented at the Annual Meeting of the American Association for the Advancement of Science.

Fisher, R. and Ury, W. (1981). *Getting to yes: negotiating agreement without giving in*. Boston: Houghton Miffin.

Greenstone, J.D. (1975). Group theories. In: F.I. Greenstein and N.W. Polsby (Eds.), *Micropolitical theory* (pp. 243-317). Reading, Mass: Addison-Wesley.

Hammond, K.R. and Adelman, L. (1976). Science, values, and human judgment. *Science, 194*, 389-396.

Holling, C.S. (Ed.). (1978). *Adaptive environmental assessment and management*. Chichester and New York: Wiley.

Howard R.A. (1975). Social decision analysis. *Proceedings of the IEEE, Special Issue on Social Systems Engineering, 63*, (pp. 359-371).

Huber, G.P. (1980). *Managerial Decision Making*. Glenview, 111: Scott, Foresman and Co.

Irland, L.C. (1975). Citizen participation - a tool for conflict management on the public lands. *Public Administration Review, 35*, 263-264.

Johnson, R.L. (1980). A multiple objective decision process for environmentally related energy development decisions. *Proceedings of the International Congress on Applied Systems Research and Cybernetics*. Acapulco, Mexico.

Keeney, R.L. and Raiffa, H. (1976). *Decisions with multiple objectives*.

New York: Wiley.

Lindblom, C.E. (1968). *The policy-making process*. New York: Prentice-Hall.

Mathes, J.C. and Stevenson, D.W. (1976). *Designing technical reports: writing for audiences in organizations*. Indianapolis: Bobbs-Merrill.

Nelkin, D. and Pollak, M. (1980). Ideology as strategy: the discourse of the anti-nuclear movement in France and Germany. *Science, Technology, and Human Values, 9*, 3-13.

Phillips, L.D. (1984). A theory of requisite decision models. *Acta Psychologica, 56*, 29-48.

Phillips, L.D. (this volume). Requisite decision modelling for technological projects.

Quinn, R.E., Rohrbaugh, J. and McGrath, M. (1985). Automated decision conferencing: how it works. *Personnel, Nov.*, 49-55.

Raiffa, H. (1968). *Decision analysis; introductory lectures on choices under uncertainty*. Reading, Mass: Addison-Wesley.

Raiffa, H. (1982). *The art and science of negotiation*. Boston: Harvard University Press.

Rapoport, A. (1970). *N-Person game theory*. Ann Arbor: University of Michigan Press.

Salisbury, R.H. (1975). Interest groups. In: F.I. Greenstein and N.W. Polsby (Eds.), *Nongovernmental politics* (pp. 171-228). Reading, Mass: Addison-Wesley.

Simon, H.A. (1976, 3rd ed.). *Administrative behavior. A study of decision-making processes in administrative organizations*. New York: The Free Press-Macmillan.

Vlek, C.A.J. (1986). Rise, decline and aftermath of the Dutch 'Societal discussion on (nuclear) energy policy' (1981-1983). In: H.A. Becker and A. Porter (Eds.), *Impact assessment today* (pp. 141-188). Utrecht: Van Arkel.

Waller, R.J. (1976, March). *Comparing structural models of complex systems*. Paper presented at the National American Institute for Decision Sciences Conference, Minneapolis, Minnesota.

Young, R.E., Becker, A.L. and Pike, K.L. (1970). *Rhetoric: discovery and change*. New York: Harcourt, Brace and World.

NOTES

1. This article is a product of the University of Michigan Value Oriented Social Decision Analysis Project (NSF Grant No. SS77-16294). We are indebted to Kenan P. Jarboe, Janet Wolfe, and Sydney Solberg for their contributions.

CONTRIBUTING TO SOCIAL DECISION METHODOLOGY: CITIZEN REPORTS ON TECHNOLOGICAL PROJECTS

PETER C. DIENEL

Citizen Participation Research Unit
Wuppertal University

1. INTRODUCTION

Our brave new overinformed world is getting ever more complex. That section of society which is expanding most rapidly probably is its steering system. There is a constant need for decision making on all levels of organization and administration (Albrecht, 1985).

Bound to the basic ideas of democracy the situation becomes even more uncomfortable: decisions have to march through the heads of many individuals. At least 51% of the relevant people in each case at stake have to consent. The decision making machinery therefore differentiates into specific institutional realms. So we find councils, political parties, specialized courts, public relation agencies, consulting firms, big science centers, and all kinds of bureaucracies, research institutes and lobbies. More and more expertise is involved. Even decision sciences are developing at a breathtaking pace. Consequently more professionalized positions emerge: the full-time social steering apparatus is growing.

This numerical growth of professionalized positions has to be controlled. There are signs of a beginning cleavage between two social classes: the privileged "decision makers" and the "administrées", the majority of the population. As documented in almost any newspaper one may look at, the typical reaction to this situation is indifference or aggression, reactions which are counterproductive to our political system. Facing the ever growing professional machinery people feel excluded from the dignity of a human existence. Saying this can not mean that we should try to reduce the number of due decisions. Mankind can not survive without exploring and developing new technologies.

The question then arises how to cope with the need of an expanding

133

Ch. Vlek and G. Cvetkovich (eds.), Social Decision Methodology for Technological Projects, 133–151.
© 1989 by Kluwer Academic Publishers.

decision making apparatus. One way of meeting the problem would be to let non-professionals participate. Why not distribute part-time opportunities for taking part in decision making to so-called lay people? The main dilemma of social decision making on technological projects seems to boil down to what has been discussed at length as the problem of public participation.

2. THE QUEST FOR AN INSTRUMENT

Scientists analyzing modern democracy agree that there is a need for broader participation (Pateman, 1970). Controversial, however, is the question of how to incorporate this into the different processes of political decision making. A number of citizens participate through the classic instruments of parliamentary democracy, like voting, like functioning in a political party, a society, a union, or even through being a member of parliament. But those instruments only offer a limited quantity of positions. One can not double or triple the number of elections, of parliaments or of governmental offices without getting negative side effects. The participatory capacity of the classic instruments proves to be exhausted (Dienel, 1988, p. 15 ff).

The same is true with respect to the supplementary techniques of citizen participation springing up these days. This is not the place to enter into a detailed discussion, but that much is obvious: plebiscitary attempts, for instance, are in danger of yielding biased results. One feels reminded of Schumpeter's saying: "The electoral mass is incapable of action other than a stampede" (1943, p. 283). To prevent this, one would have to provide the necessary broad scale information. In some cases governments may hesitate to do so. The Italian Minister of the Interior rejected to give more information before the plebiscite of Nov. 8, 1987. He didn't want "to intrude on the dialogue of the political forces". But more importantly, it generally takes an enormous input of time and money "to get a national population to know about and to participate in a policy discussion" (Vlek, 1986, p. 143). For these reasons plebiscites like that about the nuclear reactor near Zwentendorf (Austria) can be used only seldom, not often enough in the face of the number of problems to be solved today.

On the other hand vigorous information campaigning does not overcome all the difficulties. Even the intensive working with *groups* like the 'rådslag' or the 'samråd' in Sweden engaging ten thousands of people fails in certain aspects. Backed by a political party they produce onesided and therefore unacceptable results (Reinert, 1988). It is important to have these attempts but some of the techniques used succeeded only in spreading an "increased sense of needlessness of public participation" (Vlek, 1986, p. 167). So one may doubt that participation can be institutionalized along those lines. It does not seem possible to enlarge the number of people really taking part in decision making without having negative side effects on some of the situations concerned, and on society as a whole (Dienel, 1988, p. 52-64). Therefore one will still have to look for new ways to increase opportunities for participation.

Certain requirements should be taken into account in search for such a new instrument:
- *Applicability to numerous projects and people.* The new channel should not be designed for just some special cases or some "qualified persons". The citizenry at large deserves opportunities to take part in social decisions.
- *Level of information.* To offer part-time opportunities for functioning in social decisions can not mean to have more and more people decide more and more often on more and more issues about which they do not know enough. The instrument has to present the necessary information to the ordinary citizen and so provide the chance for him or her to catch up with the information advantage held by the experts. To select, to understand and to apply this information requires *time.* People making public decisions as civil servants, planners, or judges usually have this time at their disposal. The ordinary citizen, however, is expected to use leisure time to get hold of the necessary information. New modes of participation should set the citizen free from the stress of his or her daily job and should grant him or her sufficient time to carry out this kind of social responsibility.
- *Motivation to take part.* Some officially offered opportunities for taking part in public decision making remain unused. People obviously do not have the time, the money, or the kick to be present. A new instrument ought to provoke the motivation for participating in working out solutions to public problems. One way to secure this is the guaranteed seriousness of the situation. We need an instrument which can not be misunderstood as a simulation game or as just an other form of adult education. Participants should be aware that their work means business, which will not remain on paper but which is to be channeled formally into the mandatory decision procedures. Thus they will exert their influence.

- *Support of the common good.* New ways of decision making frequently work as an invitation to relevant pressure groups to step in. Everybody takes care of his or her own interest first. This is even more true for every organized interest. Crucial for social decisions, however, is the common concern (cf. Bezembinder, this volume). The structure of the instrument looked for should offer participants a chance to identify themselves with the perceivable common good. By randomly distributing the right to take part in the decision process one secures that all organized interests are excluded and included in the same way. By limiting this right to one specific case and to a short period of time the individual interests, like pursuit of a career or of re-election, may also be curbed if not neutralized. In the same way the deciding body is prevented from developing any organizational group interests. All this brings about a situation which allows participants to support the common good.

Given these requirements it would seem that the so-called planning cell (Dienel, 1980 and 1988, Garbe, 1986) is a useful way to involve the non-professionals into social decision making processes (Crosby, Kelly and Schaefer, 1986; Renn, 1986).

3. THE PLANNING CELL MODEL

The 'planning cell' is a basic element for an organized public participation, compatible with the institutionalized forms of our political/administrative system. This element consists of about twenty-five randomly selected people who are working as public consultants for a limited period of time, e.g., one week, in order to present solutions for a given planning or policy problem. The sampling process is open to persons older than 18 years of age. The people selected are invited to work and are remunerated out of public funds to compensate for their living expenses and lost earnings. The planning cell is accompanied by two process escorts, who are responsible for the information schedule and the moderation of the plenary sessions. A particular project may involve a larger or smaller number of planning cells. In each group, participants acquire and exchange information about the problem, explore and discuss possible solutions, and evaluate these in terms of desirable and undesirable consequences.

The final results of the groups' work are channeled into the institutionalized decision processes. The results are summarized as a

'Citizen Report' delivered to the authorities as well as to the participants themselves.

Organized decision making always raises the question of *representation of interests*. Formalized as well as informal decision procedures usually are highly selective: today 47% of the members of the German parliament (Bundestag) have been public servants. Only 15.4 % of the representatives happen to be women. Also with randomly composed planning cells one can not boast of perfect representation. The selected individual may accept the invitation or not. But experience shows that planning cell membership comes amazingly close to the general distribution of the social characteristics of a given population. The normal distribution indicators found in planning cells is suggested by the balance of males and females in those groups (see Figure 1).

Figure 1: Distribution of men and women in groups selected at random: the 250 people in the ten Planning Cells of the Cologne project.

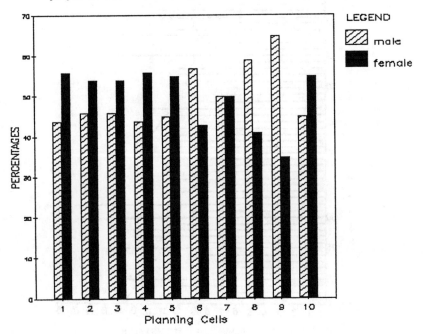

The age structure of the region from which the citizen consultants are selected is also reliably represented in the planning cells (see

Figure 2). The distribution of income brackets and status classes also looks fairly good, and the 'independence' of the people picked is acceptable too, especially if it is compared with what is common in some of the renowned public opinion research projects using sample survey methods (Hoag, 1980).

Figure 2: Distribution of age in the Planning Cell: age of the 489 public consultants in the twenty planning cells of the Citizen Report on the Future of Energy Policy.*

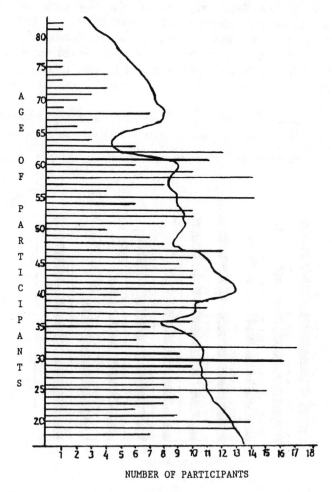

NUMBER OF PARTICIPANTS

* The underlying graph indicates the age distribution of the population of West-Germany (Statistical yearbook of the FRG, 1980).

Since the Planning Cell is designed to tackle clearly defined problems, it usually can not choose its own *tasks*. The task is determined by the authorities. Not every problem is suited to be coped with by a planning cell. On the other hand, since the issues are assigned, it is possible to propose to the groups one of the 'forgotten problems' generally disadvantaged in the existing political decision system.

For testing the new instrument only local problems were taken up like city planning (Bongardt, Dienel and Henning, 1985; Bürgergutachten, 1987), highway routing, planning recreation areas (Dienel, Friedrich and Henning, 1984), local energy supply (Bürgergutachten, 1984), or determining waste disposal sites. Since it is much more difficult to implement public participation in solving comprehensive and overlapping problems, we preferred tasks on the regional or national level during the recent years, like social decisions on energy planning (Dienel and Garbe, 1985; Peters, 1986; Renn, Stegelmann, Albrecht, Kotte and Peters, 1984; Jungermann, Pfaffenberger, Schaefer and Wild, 1986), or on new information technologies (Bürgergutachten, 1986).

The *information* relevant to the case under consideration is made available to the planning cell in different ways. The necessary technical information has to be collected and compiled by the autonomous agency in charge of running the planning cells, which so far has been the 'Forschungsstelle Bürgerbeteiligung und Planungsverfahren' at Wuppertal University.

It always appears just as necessary to consider the presentation of the positions of the various interest groups involved. This means that these positions have to be tied into the planning cell process. This can be done by on-site visits or, more often, by hearings. There the people involved present their points of view. Politicians or representatives of the administration concerned are asked to provide additional information upon request.

During much of their time the planning cell members work in *small groups* (see Figure 3). Such task-groups meet frequently during the day for screening information, for exchanging and analyzing experience, or for finding and evaluating suitable solutions. There have been, for instance, more than 1100 task-group-sessions in a middle-sized planning cell project on information technology (see

Figure 3:Group sessions during a planning cell program: mostly five groups of
         five people each.

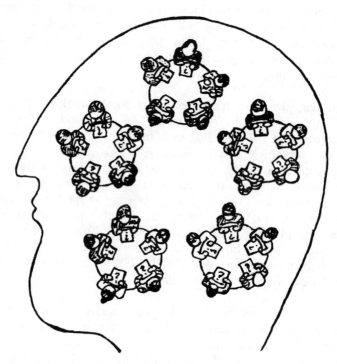

Bürgergutachten, 1986, p. 3 or p. 121). In order to prevent the
emergence of group hierarchies, membership in such small groups is
rotating from session to session.

During the small group sessions, or for summarizing the decisions
reached for further discussion, different *techniques* or decision aids
have proved to be useful. One is to integrate computer aided
dialoguing into the group-or plenary decision making processes similar
to the decision conferencing model (Phillips, this volume). In any case
the output of the planning cells is characterized by the many working
situations people go through in face-to-face speaking contacts. For
example the lay consultants of the energy project have been exposed
to each other (and to the experts) for altogether 18,633 hours of
direct communication (see Dienel and Garbe, 1985, p. 56).

In order to optimize the information and evaluation processes taking
place in planning cells, information periods, group work, hearings and

other methods have to be organized into an *agenda*. Figure 4 shows
such an agenda, as used in the "energy project".

Figure 4: A four Day's Agenda for a Planning Cell: taken from the planning cell
project "Social Compatibility of Energy Systems", 1982/83.

	TUESDAY	WEDNESDAY	THURSDAY	FRIDAY
SESSION 1	INTRODUCTION: THE ENERGY SITUATION	FOSSIL ENERGY HEATING SYSTEMS	CITY PLANNING AND ENERGY SUPPLY OR CONTINUATION OF SESSION 4 OF WEDNESDAY	GENERATION OF GOALS AND AIMS FOR ENERGY POLICIES
SESSION 2	INSPECTION TOUR OF LOCAL ENERGY FACILITY	ENERGY CONSERVATION RENEWABLE ENERGY SOURCES	INTRODUCTION TO 4 ENERGY OPTIONS* GROUP EVALUATION OF OPTIONS	FINAL EVALUATION: LOCAL ENERGY SITUATION (RECOMMEN-DATIONS)
	LUNCH	LUNCH	LUNCH	LUNCH
SESSION 3	LOCAL ENERGY SUPPLY AND DEMAND: SITUATION AND OPTIONS	NUCLEAR ENERGY	POLITICAL HEARING	INDIVIDUAL EVALUATION AND DECISION ON THE FOUR ENERGY OPTIONS
SESSION 4	INTRODUCTION OF EVALUATION CRITERIA	ELECTRICAL POWER GENERATION EVALUATION OF POWER GENERATION PLANTS		CONTINUATION OF SESSION 3 EVALUATION OF THE SEMINAR AND THE PARTICIPATORY PROCESS

* The four energy scenarios have been developed by the German Parliamentary
Enquete-Commission on nuclear energy.

Concerning the course of planning cell *projects* possibly more
questions can be raised than could be answered here. It might be

Figure 5:The Cologne Town Hall Citizen Report 1980.

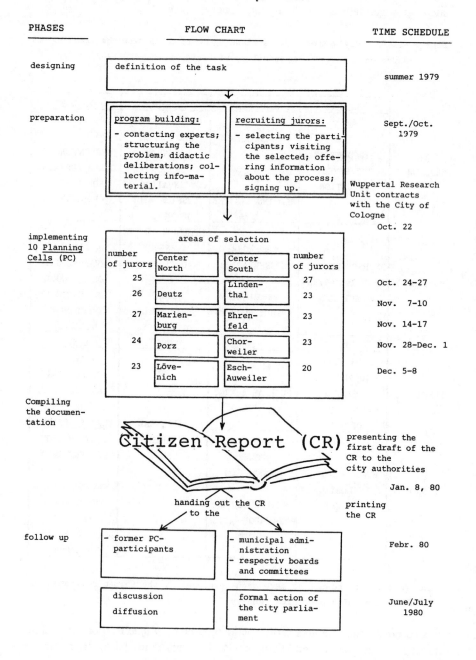

PHASES	FLOW CHART	TIME SCHEDULE
designing	definition of the task	summer 1979
preparation	**program building:** - contacting experts; structuring the problem; didactic deliberations; collecting info-material.	**recruiting jurors:** - selecting the participants; visiting the selected; offering information about the process; signing up.	Sept./Oct. 1979 Wuppertal Research Unit contracts with the City of Cologne Oct. 22
implementing 10 Planning Cells (PC)	areas of selection	

number of jurors | Center North | Center South | number of jurors
25 | | | 27 — Oct. 24-27
26 | Deutz | Linden-thal | 23 — Nov. 7-10
27 | Marien-burg | Ehren-feld | 23 — Nov. 14-17
24 | Porz | Chor-weiler | 23 — Nov. 28-Dec. 1
23 | Löve-nich | Esch-Auweiler | 20 — Dec. 5-8

Compiling the documentation

Citizen Report (CR) presenting the first draft of the CR to the city authorities

Jan. 8, 80

handing out the CR to the

printing the CR

follow up

- former PC-participants

- municipal administration
- respectiv boards and committees

Febr. 80

discussion

diffusion

formal action of the city parliament

June/July 1980

helpful to skim the flow chart describing one of these social decision projects; see Figure 5 about the Cologne Town Hall Report. The ten Cologne planning cells tried to clarify an identical set of problems. In other cases there have been up to twenty-four planning cells operating simultaneously along a standardized program for the same citizen report project. Details may be looked up in the relevant publications (see reference list).

It may be of even greater interest to have a look at the *people* taking part in these social decisions, at what it does to them, and to think about optimal methods for analyzing the results of the social decision processes and for putting them together in a final report.

4. THE INSTRUMENT 'PROCESSES' PEOPLE

The planning cell is an institutional frame for the people working cooperatively in a socially heterogeneous situation interpreted as being meaningful and rewarding. This situation does not leave people unchanged. It produces certain behavioral phenomena. They are not irrelevant to the decision output of the groups.

- Participants are learning fast; even 'hard to understand stuff' is grasped. The ability of the lay consultants to understand and to use new categories in the field of energy policies was constantly underestimated by the experts and politicians taking part (Renn, Albrecht, Kotte, Peters and stegelmann, 1985, p. 201). In the project "Solutions to Negative Consequences of New Information Technologies" people have been surrounded by hitherto unfamiliar technical equipment for four days. Working there, they acquired special knowledge with respect to this topic. Figure 6 gives an idea about the growth of participants' knowledge.
- Participants are sensitive to long-term problems, while the existing political institutions have a preference for working on problems which present themselves as "urgent" and "quickly to be solved". In the planning cells the arguing as well as some of the results reveal an unexpected sensitivity for long-term problems. Even if the task of the group is not pointing that way, the discussion insists on background issues. For example, people started to talk about Third World problems and the need for setting new terms of trade in respect to this part of the world, while working in planning cells dealing with such different topics as the evaluation of energy policy paths (Dienel and Garbe, 1985, p. 95) or the negative consequences

of new information technologies (Bürgergutachten, 1986, p. 66).

Figure 6: Distribution of technological knowledge among the participants of four
        planning cells (PC; N = 89) at the beginning and at the end of the planning
        cell in percentages.

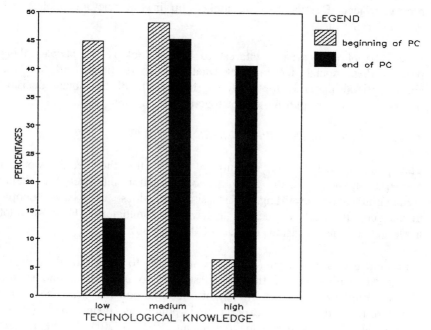

- Participants very soon feel and act like public consultants (role
  taking). They are obviously involved in their new task (Bernotat,
  1980). For example, consultants would not use their break but
  instead took their tea into the working session. From as many as
  1900 persons taking part in planning cells so far there have been
  only two or three absences.
- The reasoning of the participants is focussed upon what is
  conceivable as a common good in that specific situation (Garbe and
  Hoffmann, 1988, pp. 59-61). People take a stand for that commonly
  compatible future, even when their own individual interests may be
  violated.

5. THE OUTPUT

Every technique of public participation has to be ultimately judged by

what it achieves. The Planning Cell method yields factual results which are applicable to institutionalized decision making processes. These results are qualified by the way they are generated:
- task-oriented,
- heterogeneous with respect to values,
- sufficiently informed,
- relatively independent of special interests, and
- assessing not only present, but also future needs.

Dealing with controversial issues the output of the planning cell exercises a legitimizing effect on measures necessary in the area being planned.

Preliminary and follow-up steps (see the flowchart of Figure 5) also have their influence on the planning cell output. Its quality is highly dependent on the specific task field assigned to the group or on the resources spent on the specific case of decision making, e.g., the time assigned for information acquisition. It would be beyond our scope to describe in detail the quality of the results obtained so far. For the present purposes it seems to be more worthwhile to ask how this output is assessed and compiled.

There are different levels of decision making in planning cells: the individual person, the small group and the plenary session. On all those levels the outcomes of the debate may be collected and compared. To manifest these decisions a number of devices and techniques can be thought of. Three of them already tested should be mentioned.

5.1 Selection among presented alternatives

In the social sciences one of the oldest and most common methods to investigate opinions is to ask for preferences out of a set of given alternatives. It is used in planning cells also. At the end of the project on "Future Energy Policies" the attendees were confronted with four different scenarios ('energy paths') modeling the situation of the Federal Republic of Germany for the year 2030. These paths originally had been constructed for a German parliamentary committee, the "Enquête Commission on Nuclear Energy". They offered different options, ranging from a country with high use of nuclear energy (path 1) to one living without any nuclear energy (path 4). People were asked to select their preferred scenario. As Figure 7 shows the majority of them (53.5% select paths 3 and 4) opted in favor of a

Germany bringing down the use of nuclear energy.

Figure 7: Selection among given alternatives: final decision on the four energy
paths (in % of participating lay consultants, N = 482).

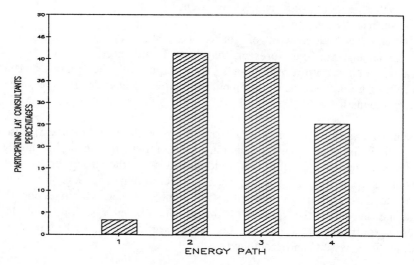

This simple decision method may also be used to obtain further
insights. The lay consultants were asked for their second preference.
Eighty-seven percent of the people originally being in favor of path
no. 2 decided to have a future without nuclear energy (see Figure 8).
But only 53% of the participants first favoring path no. 3 decided to
go up to path no. 2 (see Figure 9). Every second one of them even
voted for the extreme solution path no. 4. This kind of reaction seems
hinting at hidden tendencies of the majority of the attendees.

There have been planning cell agendas with similar evaluation
procedures taking place in all phases of their implementation. The
sequence of the completed questionnaires then represented a "skeleton
output relevant to the planning task" (Renn et al., 1984, p. 18). This
procedure makes it relatively easy to compose a citizen report, even
though it leaves some space for interpretation.

5.2 Package of decisions

Individuals or small groups may design a coordinated arrangement
(verbal or nonverbal) which integrates a multiplicity of concrete

Figure 8: Second choice among the four energy options: people who first voted in favor of path no. 2 now voted 84% in favor of path no. 3 (N=129).

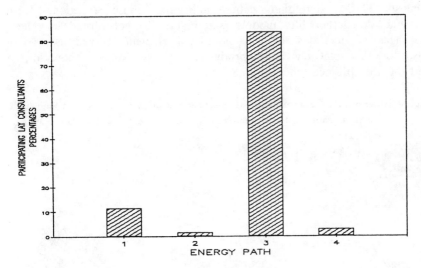

Figure 9: Second choice among four energy options: people who first voted in favor of path no. 3 now split votes (53%, 45%) between paths no. 2 and no. 4 (N=113).

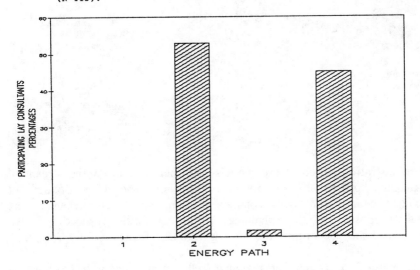

single decisions. One attempt to make use of such an output method was made at Cologne. At the end of the planning cell runs, each group of five lay consultants had to construct one model showing the distribution of seven different town functions (like flats, shops, offices) and the distribution of building units in the planned area. Almost 30,000 styrofoam cubes indicating those functions were invested. Altogether fifty models (see Figure 10) got turned in. These 'packages of decisions' using standardized elements served as a data base for the relatively quick production of the citizen report at the end of the project.

Figure 10: Arrival of some of the Cologne area planning models at the Wuppertal Citizen Participation Research Center (ed. note: the author stands at the right).

The illustrative power of the models and their display of the stream of arguments proved to be rather convincing. The City Council of Cologne being in a highly complex planning situation, it changed the final proposal of the planning authorities with respect to two strategic issues according to the suggestions of the "Bürgergutachten".

This method of presenting planning cell output may well be used in comparable cases. It is necessary, however, to realize that developing a standardized frame for such a package of decisions is rather time-consuming.

5.3 Recommendations

A third technique to manifest decisions is to conclude a project with written recommendations. They may be formulated by individuals, small groups, in plenary sessions, or by the team writing the citizen report. In that case new problems come up. As all minuting, phrasing recommendations has to follow rules. The origin of the sentences must be clearly traceable: the formulations have to be based on the quantitative analysis of the planning cells' output and later checked by the participants at a follow-up meeting.

One example to be mentioned here is the citizen report on how to handle possible negative consequences of new information technologies. This report contains seventy-eight verbal "propositions". It also presents forty-three so-called "problem indications". They describe social situations which were felt still to be unsolvable, but so important as to be necessarily considered (Bürgergutachten, 1986, p. 5-19).

6. CHALLENGES FOR RESEARCH

The planning cell instrument discussed above calls for further research in different directions. Two of them are connected with our topic:
- *Evaluating technology projects*
  New technologies exert their silent but strong influences on our society. What futures are we going to create that way? The last ten years of planning cell experience show that the ordinary people are able to understand situations that are hitherto unknown to them and that they are prepared to discuss and to evaluate. Their Citizen Reports are one way to explore the "social compatibility" of new technologies beforehand. This instrument should be adopted into research in different areas of evaluating technological projects.
- *Behavior during decision processes*
  We are interested in human decision behavior. Planning cells in a way are natural laboratory situations. Working with the standardized elements of the planning cell creates situations which are in certain respects comparable to each other (Dienel, 1988, p. 198 f.; Garbe and Hoffmann, 1988). Material is growing which documents the frames of reference, the acting persons and their development during a decision making process. This material could be sifted, analyzed and summarized against the background of

decision-theoretical frameworks.

The increased use of the planning cell and citizen report methodology will improve, as it seems, the decision making machinery of our societies. It will allow for additional societal control power and it may generate broader consensus in certain public decision areas. But above all this methodology will help an expanding social steering system in one of its crucial dilemmas by limiting the growth of a full-time apparatus and offering acceptable positions to non-professionals.

REFERENCES

Albrecht, J. (1985). *Planning as social process. The use of critical theory.* Frankfurt am Main/Bern/New York: Verlag Peter Lang.

Bernotat, I. (1980). *Die Laienplanerrolle.* Frankfurt am Main/Bern/New York: Verlag Peter Lang.

Bezembinder, Th. (this volume). Social choice theory and practice.

Bongardt, H., Dienel, P.C. and Henning, H. (1985). *Bürger planen das Rathausviertel. Planungszellen erarbeiten Ausgangsdaten für den städtebaulichen Wettbewerb Rathaus/Gürzenich-Köln.* Frankfurt am Main/Bern/New York: Peter Lang Verlag.

Bürgergutachten Energieversorgung Jüchen-Nord (1984). Forschungs- stelle Bürgerbeteiligung und Planungsverfahren, Universität Wuppertal.

Bürgergutachten Regelung Sozialer Folgen neuer Informationstech- nologien (1986). Leverkusen 3: IGEBP-Verlag.

Bürgergutachten Rathaus/Gürzenich - Köln (1987). Werkstattpapier Nr. 13. Forschungsstelle Bürgerbeteiligung und Planungsverfahren, Unversität Wuppertal.

Crosby, N., Kelly, J.M. and Schaefer, P. (1986). Citizens panels: A new approach to citizen participation. *Public Administration Review, 46,* 170-178.

Dienel, P.C. (1980). *New options for participatory democracy.* Werk- stattpapier Nr. 1. Forschungsstelle Bürgerbeteiligung und Planungsverfahren, Universität Wuppertal.

Dienel, P.C. (2nd. edition: 1988). *Die Planungszelle. Eine Alternative zur Establishment-Demokratie.* Opladen: Westdeutscher Verlag.

Dienel, P.C., Friedrich, T. and Henning, H. (1984). *Bürger planen einen Freizeitpark. Bericht über den Testlauf der Planungszelle in Soslingen.* Frankfurt am Main/Bern/New York: Verlag Peter Lang.

Dienel, P.C. and Garbe, D. (Eds.). (1985). *Zukünftige Energiepolitik. Ein Bürgergutachten.* München: High Tech Verlag.

Garbe, D. (1986). Planning cell and citizen report: a report on German experiences with new participation instruments. *European Journal of Political Research, 14*, 221-236.

Garbe, D. and Hoffmann, M. (1988). *Sociale Urteilsbildung und Einstellungs-änderung in Planungszellen.* Werkstattpapier nr. 25. Forschungsstelle Buergerbeteiligung und Planungsverfahren, Universität Wuppertal.

Hoag, W.J. (1980). Der Bekanntenkreis als Universum: Das Quotenverfahren der Shell-Studie. *Kölner Zeitschrift für Soziologie und Sozial-psychologie, 38*, 123-132.

Jungermann, H., Pfaffenberger, W., Schaefer, G.F. and Wild, W. (1986). *Die Analyse der Sozialverträglichkeit für Technologiepolitik.* München: High Tech Verlag.

Pateman, C. (1970). *Participation and democratic theory.* Cambridge: Cambridge University Press.

Peters, H.P. (1986). Social impact analysis of four energy scenarios. In: H.A. Becker and A.L. Porter (Eds.), *Methods and experiences in impact assessment.* Dordrecht/Boston/Lancaster/Tokyo: D. Reidel Publishing Company.

Phillips, L.D. (this volume). Requisite decision modelling for technological projects.

Reinert, A. (1988). *Organisierte Formen gesellschaftspolitischer Aktivierung als problem der Sozialdemokratie in Schweden.* Frankfurt am Main/Bern/New York: Verlag Peter Lang.

Renn, O. (1986). Decision analytic tools for resolving uncertainty in the energy debate. *Nuclear Engineering and Design, 93*, 167-179.

Renn O., Albrecht, G., Kotte, U., Peters, H.P. and Stegelmann, H.U. (1985). *Sozialverträgliche Energiepolitik. Ein Gutachten für die Bundesregierung.* München: High Tech Verlag.

Renn, O., Stegelmann, H.U., Albrecht, G., Kotte, U. and Peters, H.P. (1984). An empirical investigation of citizens' preferences among four energy scenarios. *Technological Forecasting and Social Change, 26*, 11-46.

Schumpeter, J.A. (1943). *Capitalism, socialism and democracy.* London: Geo. Allar & Urvin.

Vlek, C.A.J. (1986). Rise, decline and aftermath of the Dutch 'societal discussion on (nuclear) energy policy' (1981-1983). In: H.A. Becker and A. Porter (Eds.), *Impact assessment today* (pp. 141-188). Utrecht: Van Arkel.

COMPUTER ASSISTED GROUP DECISION MAKING

ROBERT DE HOOG, ERIC BREUKER
AND TIBERT VAN DIJK

Social Science Informatics Group
University of Amsterdam

1. INTRODUCTION

Large scale technological projects such as, for example, the development of nuclear energy, the storage of hazardous materials, and land reclamation, are often the subject of heated debates. Apart from matters of principle involved, this debate is frequently fuelled by the simple but often overlooked fact that the responsible decision makers rarely are also the persons who are going to suffer if something goes wrong. It would be an interesting research project to find out how many activities that are considered dangerous are actually located close to important centres of government. This seems to imply that the conflict is basically unbalanced. One party holds most of the power and will not suffer the consequences, the other is without power and runs most of the risk. From a somewhat cynical point of view one could even state that the tremendous amount of consultation that goes into some technological projects has more to do with adhering to the norms of democracy than being a real negotiation game (Munier, 1986) between more or less equal partners. This feature is succinctly caught in the notion of a social dilemma (see Wilke, this volume).

If this is true, the application of various forms of decision technology to these projects seems to be a waste of time. Most of them, in one way or an other, imply some room for compromise or trade-offs among the participants. Nevertheless the prospects of decision technology for these problems are looking rather good at the moment. Whether this is a result of the need for good justifications by the decision makers for their decisions is still an open question. The aura of 'rationality' that often surrounds this technology, will undoubtedly give decisions a less arbitrary appearance. But even when one believes in this scepticism concerning decision technology, there seems not much else to do if one wants to prevent an outright or even violent conflict between the

153

Ch. Vlek and G. Cvetkovich (eds.), Social Decision Methodology for Technological Projects, 153–172.
© *1989 by Kluwer Academic Publishers.*

parties involved. As Hober Mallow says in part one of Asimov's (1969) Foundation Trilogy: "Violence is the last resort of the incompetent". So decision technology can perhaps be used to enhance the competence of the actors.

In this contribution we will discuss a computer program which is designed for supporting negotiations and decision making in groups. We will describe the program in some detail and compare it with other computer-based methods, especially procedures presented elsewhere in this book. Furthermore we devote some attention to the applicability of the procedure to what is called here 'large-scale technological projects'.

2. A SHORT OVERVIEW OF GROUP DECISION THEORY

The purpose of this section is to give a short and necessarily incomplete overview of the more recent developments in that part of decision theory which deals with groups.

When Arrow published his famous "Social Choice and Individual Values" (1951), it became clear that no rational method was available for aggregating individual preferences into a group preference. This "impossibility theorem" constrained the research into group decision making, since no optimum method was, and will be, available. This rather severe restriction led to a somewhat different approach to problems of group decision making. Researchers became aware that the amount of information available on negotiation procedures, presented a richer field for new developments than mathematical aggregation formalisms. According to Kersten (1988) two procedures are nowadays employed for supporting group decision making: a mathematically convergent one and a psychologically convergent one. The former is based on the assumptions of rational behavior and the possibility to determine individual values that are aggregated into a group utility. The latter assumes that the decision maker is not rational or multi-rational as a group member, that there are criteria that can not be included in a model and that there are internal and external pressures present for reaching a compromise.

The most general model for (group) decision making has been formulated by Janis and Mann (1977). This model comprises four (actually five) phases:

- appraising the challenge
- surveying the alternatives
- weighing the alternatives
- deliberation and commitment
- (adhering despite negative feedback)

All decision support models will and do contain a subset or all of the phases defined above.

One of the first procedures that has been widely used to support group decision making was the Delphi-technique. Although not a proper group decision making tool, its main aim was to muster collective expertise. The element of exchanging information in order to improve decision making is still at the root of many contemporary computer programs in this field. This function has been summarized by French (1983):

"...,decision analysts should see their role as to advice the individual member of the group and to help communication and understanding between the members."

Kersten (1988) has developed a computer-based procedure where the decision makers are free to choose their own rational standards. It is capable of taking into account those aspects of the decision process which most of the time are difficult to model. Furthermore the procedure is designed to guide the decision maker in the direction (s)he prefers and in the meantime suggests the consequences of possible outcomes. The procedure is characterized by soft (the decision maker's objectives and aspirations) or strong (the negotiation setting, which is the same for all group members) constraints. Generally the procedure proceeds from a set that is strongly constrained and contains no feasible alternatives, through consecutive relaxations of the individual aspiration levels to a set with feasible options. During the iterative negotiation process alternatives are dropped from this set, until an one-element set remains.

An other approach is the one advocated by Giordano, Jacquet-Lagrèze and Shakun (1986). Their Evolutionary Systems Design (ESD) group problem representation, discerns four distinct reference spaces. These are the control or decision space, the goal space, the criteria space and the preference space. The *control space* is used for the specification of the variables used to evaluate the alternatives. In the *goal space* the various alternatives are evaluated, compared and eliminated. It is the responsibility of the users to use a consistent goal space. This might pose a problem, because the engineer can have goals that differ from those of, for example, financial experts or market forecast experts. The *criteria space* is used to reveal the reasons for conflicting positions. Its evaluations conform more or less to the format of the Delphi-approach. The *preference* or *utility space* is used to calculate for every participant their preference order. The most important goal of the program is the mapping between the four spaces. This mapping is used to present the different views of the participants to each other. We quote Giordano et al. (1986) :

> "It is important to make tools available which show to what extent participants agree or disagree, and the reasons why they are in conflict, in order to help them modify their beliefs."

An other approach is present in the POLICY program which already dates back to the seventies. This program is based on ideas from social judgment theory (Hammond, Stewart, Mumpower and Adelman, 1977). It tries to resolve cognitive conflict by means of communicating systematic differences and inconsistencies among decision makers. The program relies on regression analysis as a means of analysing individual preferences. This, however, has the drawback that a relatively large number of judgments is necessary for a reliable use of the technique. Frequently this is not feasible due to limitations in the number of realistic alternatives. A successful application of this technique is described in Steinmann, Smith, Jurdem and Hammond (1977). See further Brehmer, this volume.

Still an other example is Value Oriented Social Decision Analysis (VOSDA) implemented by Chen, Mathes, Jarboe and Wolfe (1979), see also Chen and Mathes (this volume). VOSDA performs a kind of utility analysis on the preferences of the decision makers. Because utility analysis results in decomposed preference structures, this technique seems especially suited for exchanging detailed information about each others' preferences in a collective setting. Results of the application of this procedure are reported in Chen et al. (1979) and in Chen, Mathes, Jarboe and Solberg (1981). The VOSDA procedure has a lot in common with our approach, but conducts no negotiations between the involved decision makers and/or relevant others.

Decision conferencing (Phillips, 1984; see also this volume) as such is not computer-based, but makes extensive use of different computer models like multi-attribute analysis, decision trees and resource allocation models. The main goal is to reach a "requisite model" or a "socially shared model of reality", indicating that it is not (only) the group decision that matters but also the feeling by all participants that they have reached an acceptable solution to a common problem.

Furthermore some interesting notions about formal representations of kinds of group conflicts are presented by Coombs and Avrunin (1977, see also Bezembinder, this volume). Their theoretical background can be used for giving a formal justification of the ideas underlying many of the comtemporary programs, including ours.

Both Jarke (1986) and Gray (1986) conclude that the research on computerized group decision support systems has just begun. According to Jarke (1986) any tool must meet the following criteria in order to make it potentially useful in practice:
- users should have access to the relevant data about the alternatives under review;
- the users can change their minds, interpretations of data will differ over time, this must be accommodated;
- users have different areas of knowledge, the program should facilitate the sharing of knowledge among users;
- users always have different points of view, programs should provide tools for tracing inconsistencies in the cognition of users, furthermore the negotiation process must be supported as well; and
- users must be able to employ secret rules and data.

The first four points are more or less met by the programs discussed, though not equally well. In particular the first, data base, point, is seldom satisfied, due to an enduring wide gap between data base technology and decision technology. The last point is definitely not in line with the openness of preferences and opinions propagated by all programs. Consequently none of them supports this facility.

We may conclude that it is the main purpose of all programs to support decision making and canalize the negotiation process, and not to decide which alternative has to be chosen. Exchange of information is the key to the two main purposes. An other quite common feature of the programs is the willingness of the participants to compromise, to water down their claims. If this is not the case a deadlock will occur, and no program will ever be able to break these deadlocks without changing the minds of the people involved. As we will explain in the next section, our PANIC program also falls in the main category of programs discussed in this section.

3. THE PROGRAM PANIC[1]

From the previous section it is clear that computer assisted group decision making is not a completely new phenomenon. A number of programs do exist already, albeit mainly in the United States. A fairly comprehensive overview of these is also presented by Gray (1986). We will limit ourselves in this contribution to a comparison of our program with other procedures discussed in this volume. Before such a comparison can be made, the PANIC program must be described.

PANIC is the acronym for Program Assisting Negotiations In Collectivities. This name already points to one of its aims: supporting a negotiation process in which two or more people take part. The basic idea behind the program is that negotiations are often a process of give and take by individuals. By focusing the discussion and negotiations on the relevant issues, the program tries to speed up decision making. At the same time the program can be used to identify differences in knowledge between group members as well as unbridgeable gaps between the values of the participants. Furthermore we want to emphasize that the program is not meant to take over the decision process. At every moment in the procedure the group can quit or pass over its advices.

Because PANIC contains an other program for the individual part of the procedure and the latter program defines the basic structure of the negotiation process, we will first give a condensed description of that program in the next section.

3.1 The MIDAS program

This Multi-attribute Individual Decision Assistance System tries to support individual decision making by offering a procedural model of how a decision process can be organized. The underlying model conforms to the ideas embodied in multi-attribute utility analysis (see, for example, Von Winterfeldt and Fischer, 1975). Programs from the same strand are MAUD (Humphreys and Wooler, 1981) and Déja Vu (Intelligent Environments Inc.). During the last five years we have

conducted a fair amount of research concerning the usefulness of the
MIDAS program for different decision domains and different kinds of
users (Bronner and De Hoog, 1983; Bronner and De Hoog, 1984).
Currently a project on the applicability of MIDAS to the choice among
university studies is carried out at the University of Leyden (Bakker,
Topman and De Hoog, 1984, provide a description).

The analysis incorporated in MIDAS, guides the user through eleven
stages. As we assume that most of our readers are familiar with multi-
attribute utility analysis, we will explain each phase only cursorily
below:

(1) *specifying the domain*: the user indicates the domain of the
    decision in order to focus the phrasing of sentences (for example
    'cars', 'houses', etc.);

(2) *input of the alternatives*: the user supplies the names of the
    alternatives (s)he wants to consider;

(3) *eliciting attributes*: the user states the attributes (s)he sees as
    relevant for the decision in the domain. These attributes may
    either be generated by means of Kelly's Repertory Grid method
    (Shaw, 1979), by a direct naming of the attributes, or by
    selecting a number of them from a prepared list (if the domain is
    known beforehand);

(4) *location measures*: the user assesses the location of each
    alternative on each elicited attribute. This location measure is
    either based on the knowledge of the user or it may be retrieved
    from a suitably designed data base accessible from the program;

(5) *ideal points*: on each attribute the user gives his most preferred
    position ('ideal point') which can but need not be located at one
    of the extremes of the attribute;

(6) *redundancy checks*: the program tries to find out whether the
    structure that has been built in the previous steps does not
    contain redundancies: are attributes conceptually clear, do
    attributes discriminate sufficiently between the alternatives?
    Redundancies may be removed by the user.

(7) *check for dominance*: the built structure may contain a dominant
    alternative, i.e., an alternative outperforming all alternatives on
    all attributes. This cuts the program short, because that
    alternative obviously is to be preferred;

(8) *weighting*: the user may indicate differences in importance
    between the attributes. Currently two methods for doing this are
    available: magnitude estimation and direct (alphanumeric) scaling;

(9) *suggestion*: the program presents to the user a preference order of the alternatives under consideration. This order is based on a calculation performed on the data put in so far;

(10) *explanation*: the program may give the user an, at this moment, not very sophisticated explanation of the value that determines the position of an alternative in the preference order.

(11) *sensitivity analysis*: if the user feels dissatisfied with the suggestion of the program or wishes to test the 'stability' of its advice under small changes in the input data, a sensitivity analysis may be performed on location measures, ideal points and weights (not all fully implemented yet).

From research mentioned before it has become clear that MIDAS is a useful tool for guiding individual decision making. One of its strongest contributions, however, is to something called 'consciousness raising' (Humphreys and McFadden, 1980). This means that getting insight into the structure and nature of one's own preferences is an important side-effect of using the MIDAS program. Users frequently reported this effect as having more value for them than the program's advice on which alternative to choose. This seems to fit well into the view of MIDAS as an assistant for clarifying individual preferences. The PANIC program tries to take advantage of this effect when communicating decomposed individual preferences to all group members, making them aware of the state of mind of the others. This was the main reason to incorporate the greater part of MIDAS in PANIC. Apart from that, the individual data supplied by each group member are input for PANIC.

3.2 The elements of PANIC

Just as the program described in the previous section, PANIC also consists of a number of phases, each with its own goals and procedures. These phases will be described in the next subsections. Due to limitations of space we are not able to present all the details of the algorithms.

3.2.1 The pre-structuring phase

In order to keep the process manageable it is necessary that prior to the start of PANIC the group reaches agreement on the viable alternatives that will be the subject of decision making. This does not mean that later on no new alternatives can be put forward. The limitation holds only during the operation of PANIC. Furthermore the group must generate a set of well defined attributes relevant for the decision. This set can be obtained by including all attributes suggested by members of the group. In practice this attribute set will need some editing, because some attributes may appear repeatedly under different names, others are ambiguous, and for some no reliable location measures or positions of alternatives on them are imaginable. This prepatory work is outside the scope of the program and must be performed in a way agreed upon by all members. The result is the so-called *initial structure*, the starting point for the next phase.

3.2.2 The data-input phase

The aim of this phase is entering the data agreed upon and edited by a responsible person into a common data base. This is a rather simple part of the procedure, which can also be performed by a responsible person. This responsible person is either a member of the group or an outsider asked to do this task.

In some cases, notably when there are different levels of knowledge among the group members, the responsible person, being an expert (him)herself or having easy access to the necessary expertise, can assign positions (location measures) to alternatives on attributes. These positions may then be consulted by the less knowledgeable group members; this prevents unnecessary guesswork.

3.2.3 Individual analysis

Each member of the group works through a decision supporting computer program that is very similar to the MIDAS program described in the previous section. Some elements of that program are superfluous, new elements are added. Of course there is no need to elicit attributes, because these are already available from the previous phase. Each member can choose from these attributes the ones (s)he wants to incorporate in his or her decision process. As a result every member can define his or her own 'decision space'. The most important extension has to do with the negotiation facility of PANIC. It is the assumption underlying PANIC that negotiations must involve only the affective elements in the decision structure of each individual. In other words: you can negotiate about values, not about facts. The value side of decision making manifests itself in the ideal points and in the attribute weights. In the present implementation of PANIC the negotiation procedure is focused on weights.

Essentially PANIC tries to promote consensus by persuading group members to be a bit lenient about their weights. By being prepared to give somewhat less weight to some attributes and somewhat more to others, each group member can contribute to reaching consensus. The rationale behind this procedure is the common sense notion that in order to reach agreement one sometimes has to come down a peg or two. This notion is operationalized in PANIC by so called *negotiation intervals*. Each group member not only has to give weights to all attributes, but must also indicate for every attribute-weight combination, an interval within which the actual weight may fluctuate. These intervals define each individual's negotiation space. The negotiation spaces of all members constrain the kind of consensus that can be reached.

3.2.4 Analysing differences of opinion

As has already been said, the negotiation procedure is focused on the value side. As a consequence, one must first attempt to resolve differences of opinion before one can proceed. That is why in this phase all alternative-attribute positions are analysed for all group members. The program calculates the median position of each alternative on each attribute. After that it searches for group members who have assigned positions that differ significantly from the medians. Those members may reconsider their positions, and if necessary the group can discuss together what would be 'correct'. If no agreement can be reached, the conclusion must be that there is uncertainty about these positions. In such cases every member may stick to his/her positions.

This separation of fact and value is a difficult but very important part of the procedure. It is therefore essential that in case of disagreement the group can discuss differences. This is an important tool for uncovering arguments in which lack of knowledge is disguised by values. This disguise may either be subconscious or deliberate. The possibility that one has to defend one's positions in public is essential. Without it the procedure becomes liable to manipulation or even outright fraud.

3.2.5 Consensus-seeking phase

This phase has two possible outcomes: consensus is reached or it is not. First, the program identifies two alternatives that are good candidates for consensus. This is not strictly necessary because the calculations can be performed for all alternatives, but it speeds things up considerably. These two alternatives are called the *Most Likely First (MLF)* and *Most Likely Second (MLS)* alternatives. Passing over details of that procedure, the next step is finding out whether the group can reach agreement about either the MLF or the MLS. This will be true whenever each group member has the MLF or the MLS at rank one or two in his/her individual preference order. In other cases the program tries to reach this situation by changing the weights of the attributes within the intervals indicated, of course only for those members that who do not have MLS or MLF at rank one or two. In other words: the program uses each relevant member's negotiation space (partly or wholly) in trying to reach consensus. After changing the weights it recalculates the preference order. If after these calculations all group members have MLF or MLS at rank one or two in their individual preference rank order, consensus is atttainable. It is up to the group to follow the advice of the program: they can either choose MLF or MLS, or decide to try again.

An important thing to note in this procedure is that nowhere individual values are aggregated, so it does not violate constraints on interpersonal utility comparisons (cf. Bezembinder, this volume). The whole procedure is carried out within the boundaries indicated by each individual. To our opinion this is an attractive property of the program because it essentially leaves control to each member. A possible negative effect is that the procedure may be sabotaged by members. The obvious way to do this is by giving very tight negotiation intervals. But because all individual preference structures will be communicated to all members, this strategy will be uncovered. The only possible conclusion in such cases is that the individual does not want to negotiate at all. This is of course his or her right and it can be expressed through the program. One has to concede however that in such situations all produres, short of coercion, will fail.

If no consensus can be reached the program goes back to the facts once more. Whenever there are differences of opinion concerning the facts left, the program tries to iron out those differences. We will not give the details of this procedure. It suffices to state that the rationale behind this part lies in the apparent uncertainty about the facts. It comes close to a standard sensitivity analysis in which one tries to determine what the effects on preferences are of (small) variations in the relevant data.

Not reaching consensus in both variants indicates that the problem is outside the solution scope of PANIC. Most likely there are significant discrepancies between the group members and it is unlikely that, given the alternatives and attributes, consensus is possible at all. The experience with PANIC, however, may be used to start a new cycle, because most of the time more knowledge will be available to all members. This knowledge can be used for trying to find new alternatives that are more acceptable or for obtaining more information about the alternatives.

As has already been said, a lot of output is generated in the process. The main goal of the output is to communicate details of preferences among group members. Furthermore this output is also necessary for preventing misuse of the procedure, making possible the detection of manipulation of facts or weights.

3.2.6 Group meeting

As most of the negotiations have been conducted via the program, the group meeting can be very efficient and effective. The content and procedure of the meeting depends of course on the results of the different phases. If consensus is straightforward, no problems will arise. If weights have been changed by the program, some members may object to the consensus proposed. This may be based on the argument that, when giving negotiation intervals, they could not gauge the possible effects. This might be a reason for starting a new procedure. Whenever consensus is not attainable, the output of PANIC may be used for planning new steps.

3.2.7 Summary of PANIC

It must be emphasized that PANIC plays an advisory role throughout. Its contribution to the group decision making process can be summarized in the following points:

- structuring and making explicit the preferences of all members;
- as a result those preferences become transparant and can be talked about;
- by eliciting individual negotiation spaces the opportunities and constraints in the negotiation process are made explicit;
- by determining whether consensus is attainable a lot of time can be gained by foregoing fruitless discussions about matters that

would not influence the result at all;
- in trying to separate facts and values, discussions are kept to the point and (fundamental) uncertainty may be detected;
- the procedure may act as an eye-opener for indicating new ways to restructure or redefine the decision problem in those cases where no satisfactory consensus is reached either by the program, or because the members disagree with the program's advice.

This does not mean that the program is a panacea for all group ills. It has very definite limits, which are, by the way, almost the same for all computerized decision support systems. Data about the practical application of PANIC in groups may be found in Breuker, Van Dijk and De Hoog (1988a, b).

3.3 A comparison

In this volume three other procedures for supporting group decision making are described (by Brehmer, Chen and Mathes, and Phillips, respectively). These are summarized in Section 2 of this contribution. The question is what the differences and common elements among these are. Unfortunately this question is not easy to answer without repeating much of the details of the procedures, something which is not feasible within the limits of this chapter. In order to structure matters somewhat we will conduct the comparison from the angle of PANIC.

First we note that only PANIC supports a negotiation process by carrying out calculations on the negotiation intervals. This facility furthermore requires from the participants that they express their negotiation space in advance. We do not want to claim that this facility is an asset in all situations, but if it is not, it may be left out very easily.

Second, we think that our procedure has most in common with VOSDA (Chen and Mathes), because it also relies on multi-attribute utility analysis. Apart from the negotiation aspect, the main difference lies in the kind of utility model employed. We use a utility model including individual aspiration levels ('ideal points') and single-peaked preference functions, while VOSDA works with the straightforward additive utility model. This is more than a superficial difference, because including aspiration levels offers the possibility to get insight into an other aspect of each participant's preference structure.

Compared with POLICY (Brehmer) the main difference lies in the kind of model used (regression vs. utility analysis). This shows itself in a radically different approach to treating judgments. POLICY works with holistic judgments about composite alternatives. On the contrary our approach, and VOSDA also, decomposes the alternatives first into a multi-attribute representation, inferring the expected holistic judgment from the decomposed structure according to the utility model used. Which approach is to be preferred depends on the situation. Our approach can not put more into the decomposed structure than the participants are willing to introduce. POLICY, on the other hand, makes it possible that unexpected facets come to the surface by analysing the holistic judgments outside the control of the judge him- or herself. It seems that POLICY is especially suited for those cases in which the decision domain is fuzzy, requiring an initial structuring step. A disadvantage of POLICY is its need for a rather large number of judgments (and, by consequence, of alternatives) in order to perform a reliable regression analysis. This condition may lead to difficulties in practice.

Decision conferencing (Phillips) seems to be most different from our approach. This is, however, only partly true, because during decision conferencing multi-attribute models are sometimes used. From reports thus far (e.g., Quinn, Rohrbaugh and McGrath, 1985) it appears that most of the time a valuable contribution comes from resource allocation models. Furthermore this procedure depends heavily on an intensive face-to-face meeting of the participants involved. It also includes some notions from group dynamics, which are conspicuously absent in the other methods. Because no complete description of the procedure(s) is available, a more detailed comparison is not possible at the moment. However, to our judgment decision conferencing seems to work best when the participants have a (very) strong incentive to reach a solution for a common decision problem. This incentive is often related to being a partner in the same organisation whose survival will in the long run benefit all participants, albeit in different amounts.

4. PANIC AND DECISION MAKING ABOUT LARGE SCALE TECHNOLOGICAL PROJECTS

In the Introduction we discussed the apparently unbalanced conflict that seems to characterize most of the controversies surrounding more or less hazardous or environment-threatening projects. Could procedures like PANIC help under such circumstances?

Relatively small-scale (organisational) decision making more or less meets the characteristics summed up by Gray (1986):
  - decision making is a joint activity, engaged in by a group of people of equal or near equal status, typically involving five to twenty individuals;
  - the activity and its outputs are intellectual in nature;
  - the product depends in an essential way on the knowledge, opinions and judgments of the participants;
  - differences in opinion are either settled by fiat by the ranking person, or more frequently by negotiation or arbitration.

In contrast, key charateristics of *large-scale technologies* (LST) projects, as proposed by Vlek and Cvetkovich (this volume, 'Introduction and Overview' chapter) are:
  - LST projects can be understood, evaluated and controlled by few, but their consequences may affect very many people;
  - they are aimed at massive (and immediate) benefits, but may also have disastrous (remote) consequences;
  - the various affected groups may have divergent positive or negative interests in the consequences of an LST project;
  - the novelty and complexity of many LST projects make their possible consequences and effects very hard to assess, and the projects themselves difficult to control.

A comparison of these two lists shows quite clearly that they do not have much in common. The fact that most of the decision support programs described in Section 2 (including PANIC) are geared towards the first type of decision context, could be a reason to discard PANIC and also the other procedures discussed in this volume. Decision conferencing, for example, seems to make sense only when all participants are more or less dedicated to a common cause, as is often the case in organisational settings. A "socially shared model of reality" (Phillips) can only be an aid to improved decision making if (nearly) everyone has something to gain from "sharing" his or her model.

The main point that emerges seems to be, that vis à vis large-scale technological decision problems that incorporate real social dilemmas (see Wilke, this volume), decision technology offers only very modest tools. This holds *a fortiori* for PANIC. It does not mean, however, that the program can have no value at all. Admittedly, the negotiation component seems to be the least useful one in those situations in which one of the parties has not much to trade for. Things like "negotiation intervals" and "most likely first alternative" are more or less meaningless in those contexts where one knows in advance that preferences will be diametrically opposed to each other. Furthermore, most procedures are based on the idea that some kind of compromise is possible. Stated in the language of decision theory: one can trade one aspect of alternatives for an other.

In the case of choosing among cars, for example, a group member may be willing to sacrifice some economy in order to gain some speed, resulting in a slight change in the kind of car preferred. Unfortunately, most decisions concerning large-scale technological projects do not exhibit this feature. In a country like The Netherlands the continuing discussion between proponents and opponents of nuclear energy can not be resolved by building, for example, only half the proposed amount of nuclear power plants, because 'Chernobyl' taught us that the effects of a major accident in one plant may be quite destructive to a country of this size. The same holds for land reclamation in the large Markerlake (north-east of Amsterdam): only a half Markerwaardpolder will satisfy neither the pro-Markerlake activists nor the Markerwaardpolder lobby. This indivisibility of choice alternatives forms a considerable barrier to using existing group decision technologies.

Nevertheless in these situations the communication of opinions may be valuable, because it gives at least some insight into the details of the preferences of the other parties. This would force that party to defend its opinions, if necessary in public. The separation of fact and value, advocated by PANIC, is at the same time the most difficult and most promising part of the procedure. Most difficult, because as Brehmer (this volume) states, there is not only disagreement with respect to values, but also about facts. This will be quite common as a rule.

Very frequently facts and values get intertwined and mixed up, confusing discussions because one party talks about "facts and evaluation of facts", while the other wants to talk about facts or evaluations separately. By trying to isolate the factual elements of decisions, the differences between experts and other people involved can be made explicit. If there is not sufficient knowledge about a certain fact, the PANIC procedure will promote a discussion that will expose this lack of knowledge, thereby preventing that the cloak of evaluation is covering the absence of relative certainty.

It seems to us that (parts of) PANIC may be used by the less powerful and perhaps less knowledgeable participants as a tool to challenge the authorities to defend their position in a very explicit way, perhaps uncovering where facts give way to values. Although using something like PANIC in a nation-wide discussion seems far out, the tool might be used by special interest groups in relatively small meetings with authorities. During these meetings the authorities have to make clear the structure of their preferences by assigning positions, giving ideal points and giving weights to attributes. This structure may then serve as a focus for subsequent discussion and at the same time as a kind of "analytical knife" for dissecting opinions. Experiences with the MIDAS program mentioned in Section 3, convinced us that decomposing someone's preferences is often a prerequisite for making them understandable at all.

Of course there is no guarantee that this will work; argumentation is definitely not one of the main assets of computer programs. That aspect will rightly remain in the realm of politics and the relative persuasiveness of the parties involved. But if this modest enhancement of competence does not work, the limits of decision technology, whatever its form, will have been reached.

5. CONCLUSION

Computer assisted decision making seems to be caught in a dilemma, which could well be the dilemma of all support systems. On the one hand, one wants to assist and improve the characteristics of the decision process in order to reach "better" decisions. On the other hand, this noble purpose implies imposing a structure on the same process, a structure which may be contrary to the wishes of the participants. So in designing these systems one has to sail between the Scylla of imposing too much structure and antagonizing the parties, and the Charybdis of imposing too weak a structure and not helping them at all. The obvious solution is an "intermediate" one, but no one seems to know what "intermediate" in this case means.

We believe that the PANIC procedure is rather close to this intermediate solution. The reasons for this are threefold:

- due to the fact that all participants can stay in principle in their own decision space, defined by the attributes selected and negotiation intervals assigned, they are not 'averaged out' outside their control;
- the procedure promotes discussion about relevant differences in knowledge and information levels; no 'knowledge' can be declared dominant by one of the participants (or, for example, by an outside source of information);
- the program *suggests* an acceptable alternative to the group; we would never advise a group to rely for their decisions on the advice of a computer program only.

As a result we think that the relative *autonomy* of the decision makers is guarded as well as possible in this procedure. Whether this is true remains of course a topic for empirical research.

This does not mean that PANIC as a tool has reached its definitive form by now. A number of improvements are being carried out at the moment. These are, among others: making it easier to ask 'what if' questions, including individual loss functions indicating how much each individual will 'lose' with the proposed consensus, and implementing a backward search strategy making it possible to advise the group about the best available solution for a problem. Beside these, one of the main bottlenecks remains the elicitation of adequate attributes. At the moment no satisfactory procedure is available for this within MIDAS or PANIC. Perhaps some contribution to this problem will come from the field of artificial intelligence and knowledge representation.

REFERENCES

Arrow, K. (1951). *Social choice and individual values*. Cowles Commission Monographs, 12. New York: Wiley.
Asimov, I. (1969). *Foundation*. Panther Books.
Bakker, J., Topman, R. and De Hoog, R. (1984). *Studiekeuze en computerbegeleiding*. Paper presented at the Dutch Psychology Conference, Ede.
Bezembinder, Th. (this volume). Social choice theory and practice.
Brehmer, B. (this volume). Cognitive dimensions of conflict over new technology.
Breuker, E., Van Dijk, T. and De Hoog, R. (1988a). Can a computer aid group decision making? In: B. Munier and M.F. Shakun (Eds.), *Compromise, negotiation and group decision*. Dordrecht: Reidel.

Breuker, E., Van Dijk, T. and De Hoog, R. (1988b). *Strengths and weaknesses of computerized group decision support*. Paper presented at the FUR-IV Conference, Budapest.

Bronner, A.E. and De Hoog, R. (1983). Non-expert use of a computerized decision aid. In: P. Humphreys, O. Svenson and A. Vári (Eds.), *Analyzing and aiding decision processes* (pp. 281-299). Amsterdam: North-Holland.

Bronner, A.E. and De Hoog, R. (1984). The intertwining of information search and decision aiding. *Acta Psychologica, 56*, 125-139.

Chen, K. and Mathes, J.C. (this volume). Value oriented social decision analysis: a communication tool for public decision making on technological projects.

Chen, K., Mathes, J.C., Jarboe, K. P. and Solberg, S. (1981). Alternative Energy Futures: Interest Group Tradeoffs. *Proceedings of the International Conference on Cybernetics and Society* (pp. 548-552). New York: IEEE.

Chen, K., Mathes, J.C., Jarboe, K. and Wolfe, J. (1979). Value oriented social decision analysis: enhancing mutual understanding to resolve public policy issues. *IEEE Transactions on Systems, Man and Cybernetics, 9*, 567-580.

Coombs, C.H. and Avrunin, G.S. (1977). Single peaked functions and the theory of preference. *Psychological Review, 84*, 216-230.

Déja VU (1986). *PC World, April*, 142-148.

French, S. (1983). A survey and interpretation of multi-attribute utility theory. In: S. French, R. Hartley, L.C. Thomas and D.J. White (Eds.), *Multi-objective decision making* (pp. 263-277). London : Academic Press.

Giordano, J.L., Jacquet-Lagrèze, E. and Shakun, M.F. (1986). *A decision support system for design and negotiation of new products*. Paper presented at the FUR-III Conference, Aix-en-Provence.

Gray, P. (1986). Group decision support systems. In: E.R. McLean and H.G. Sol (Eds.), *Decision support systems: a decade in perspective* (pp. 157-173). North-Holland.

Hammond, K.R., Stewart, T., Mumpower, J. and Adelman, L. (1977). Social Judgment Theory. In: M. Kaplan and S. Schwartz (Eds.), *Human judgment and decision processes*. New York: Academic Press.

Humphreys, P. and McFadden, W. (1980). Experiences with MAUD: aiding decision structuring versus bootstrapping the decision maker. *Acta Psychologica, 45*, 51-69.

Humphreys, P. and Wooler, S. (1981). *Development of MAUD4*. Decision Analysis Unit, Technical Report 81-4. Uxbridge, Middlesex: Brunel

University.

Janis, I.L. and Mann, L. (1977). *Decision making; a psychological analysis of conflict, choice and commitment*. New York: The Free Press.

Jarke, M. (1986). Group Decision Support through Office Systems: Developments in Distributed DSS Technology. In: E.R. McLean and H.G. Sol (Eds.), *Decision support systems: a decade in perspective* (pp. 145-157). North-Holland.

Kersten, G.E. (1988). Generating and editing compromise proposals for negotiations. In: B. Munier and M.F. Shakun (Eds.), *Compromise, Negotiation and Group Decision*. Dordrecht: Reidel.

Munier, B. (1986). *Negotiation games with MCDM*. Paper presented at the 23rd Meeting of the Euro Working Group on Multiple Criteria Decision Support, Rotterdam.

Quinn, R.E., Rohrbaugh, J. and McGrath, M. (1985). Automated decision conferencing: how it works. *Personnel, Nov.*, 49-55.

Phillips, L.D. (1984). A theory of requisite decision models. *Acta Psychologica, 56*, 29-48.

Phillips, L.D. (this volume). Requisite decision modelling for technological projects.

Shaw, M.L. (1979). *On becoming a personal scientist*. New York: Academic Press.

Steinmann, D.O., Smith, T.H., Jurdem, L.G. and Hammond, K.R. (1977). Application and evaluation of social judgment theory in policy formation: an example. *Journal of Applied Behavioral Science, 13*, 69-88.

Von Winterfeldt, D. and Fischer, G.W. (1975). Multi-attribute utility theory: models and assessment procedures. In: D. Wendt and C.A.J. Vlek (Eds.), *Utility, Probability, and Human Decision Making* (pp. 47-87). Dordrecht: Reidel.

Wilke, H.A.M. (this volume). Promoting personal decisions supporting the achievement of risky public goods.

NOTES

1. The MIDAS program is written ISO-standard PASCAL and runs under UNIX(R) and MS-DOS operating systems. The program (in Dutch and in English) can be obtained from the first author. The PANIC program is written in a DBASE-III compiler version (CLIPPER) and runs on any MS-DOS system with a hard-disk. The program (English and Dutch versions) can be obtained from Delphi Software B.V., Stadionkade 134-II, Amsterdam.

TECHNOLOGICAL RISK, POLICY THEORIES AND PUBLIC PERCEPTION IN CONNECTION WITH THE SITING OF HAZARDOUS FACILITIES[1]

MATTHIJS HISSCHEMÖLLER AND CEES J.H. MIDDEN

Department of Sociology
Erasmus University Rotterdam

Department of Psychology
University of Leiden

1. INTRODUCTION

This chapter deals with the relation between public policy making and public opposition in the case of siting hazardous technologies, especially noxious wastes and nuclear power plants. Our main purpose is to show that the realisation of a site - or, what might be considered to be the opposite, its prevention by local residents and environmental protection groups - by no means solely depends on the way 'the public' perceives risk and is decided to resist a facility. By and large the outcome of a siting process is determined by the way governmental agencies perceive social reality and behave according to their perceptions. Different siting strategies used by national, regional and local authorities are not seldom based on different sets of assumptions, here called *policy-theories* (Leeuw, 1986; Van de Vall and Ulrich, 1986).

Our line of argument is as follows: first we explore, in Section 2, the various assumptions underlying public and governmental behavior. We distinguish four policy-theories concerning the 'siting problem'. Each of these can be founded in normative political philosophy. In Section 3 we point to some relevant findings of public perception research. We briefly compare these findings with the implications of the policy approaches to the siting problem. Next we examine in more detail how the various siting approaches and public perception of risk may become interwoven in siting practice. Therefore, in Section 4, we compare five siting strategies in different political systems. Finally, in Section 5, we identify three *general* conditions explaining when and why different strategies may be effective.

Ch. Vlek and G. Cvetkovich (eds.), Social Decision Methodology for Technological Projects, 173–194.
© *1989 by Kluwer Academic Publishers.*

2. FOUR POLICY THEORIES OF THE SITING PROBLEM

Policy research in the field of siting has largely concentrated on (the absence of) adequate decision making procedures. In order to contribute to the solution of the siting problem new policy instruments have been proposed, especially in the sphere of incentives and compensation (Carnes, Copehaver et al., 1983; O'Hare, Bacow and Sanderson, 1983; Paige and Owens, 1983). In this chapter we will follow an other orientation. The appropriate method, in our view, is to reconstruct the policy theories, e.g., the implicit notions on the siting problem that underly the development and application of policy instruments. In most countries not only a variety of instruments is applicable, but different instruments may also provide different kinds of use.

We suggest that - at least in the case of siting hazardous technologies - there are four major policy theories for dealing with the problem from a public policy point of view: we distinguish a *technical*, a *public participation*, a *market* and a *distributive justice* approach. In real policy processes these policy theories - or, approaches to the siting problem - often appear in mixed combinations. The reason for this is that each of them stresses only one aspect of democratic rule and none of them can be ignored completely. We shall show how these approaches are partly complementary and partly contradictory.

2.1 The technical approach

Local residents' reactions to siting proposals are often considered to be driven solely or almost exclusively by emotions and fears. Some argue that public emotions and fears are 'natural', as laymen are not capable to understand technical processes and measures to prevent harm for health and the environment. This incapability provides the main cause for emotional reactions and prevents people from judging rationally. Rationality in the technical approach is thus primarily considered to be an effect of knowledge. Following this argument emotions and fears become the opposite of rational behavior (cf. Van der Pligt, this volume). Hence, this rationality might also imply that, just because one is aware of lacking specific knowledge, one relies upon experts who are rational by their very profession.

From this view, public resistance against a hazardous waste facility seems almost inevitable. Some refer in this respect to experiences in the past (see Edwards and Von Winterfeldt, 1985); for instance the resistance against the introduction of railways in the 19th century. These experiences show, after all, public opposition to disappear as people get used to the new technology. If a siting strategy would be based on the technical approach exclusively, it would certainly contain the following elements:

1. There is no need for delay because of public opposition. Public participation in technological decisions is undesirable, though information transfer may mitigate severe opposition.
2. A rational political process implies that politicians rely upon experts. Interventions of elected political leaders in matters considered 'technical' may even disturb efficient decision making.
3. The technical approach is not able to cope with disagreements among experts themselves. In countries like the Netherlands and Sweden, where the scientific community is rather small, cases of environmental policy often reveal a single-minded scientific community that hardly tolerates divergencies of opinion.

The elements just mentioned may look very familiar to those who have observed the field of nuclear energy policy in the U.S. and Europe during the fifties and sixties. But at the time nuclear policy started to be questioned, the technical approach began to loose its dominant position in the policy process. Not surprisingly this happened when environmental problems were recognized and large-scale technologies like nuclear energy came under attack (Zinberg and Deese, 1980; Zijlstra, 1982). At that very moment problems merely considered as 'technical' became 'political' problems and, together with the consensus about the principles of techno-economic progress, the technical approach was forced to share its position with other, contradictory approaches, especially the one favouring public participation. Three events constituted the main cause of its decline:

1. Experts proved to disagree on key-elements of fission technology, especially the solution of the radioactive waste problem.
2. Resistance against nuclear energy evolved largely, in some countries culminating into an organized public debate about nuclear energy and the waste problem.
3. Recently, awareness about the hazardous waste problem in general increases in all industrial countries. Administrative procedures have proved inadequate to prevent illegal or semi-legal dumping of hazardous waste in urban and rural areas.

2.2 The public participation approach

There are many options about the rights people have to resist governmental decisions, individually as well as collectively. These options differ with respect to the emphasis put on direct or indirect democracy. Forms of *direct* democracy are referenda (e.g., in Sweden) and free negotiations. More 'paternalistic' (Susskind and Elliott, 1983) procedures for *indirect* involvement are consultative referenda (e.g., in France), public hearings (e.g., in the U.S.A.) and public inquiries (e.g., in the U.K.). Despite these differences, all supporters of participation in siting decisions start from the assumption that people are capable of making rational judgements on these matters (which, however, does not mean they always do).

To defend their rights people organize themselves in local safety committees that negotiate with public officials, start judicial procedures, inform the press, call for a referendum, etc. These groups claim to represent the majority of those mostly involved. They also claim that they can provide information necessary to make the best decision. In many cases local and environmental protection groups invoke experts' support in order to evaluate the information provided by governmental experts.

The 'best' decision is argued to be 'just' in two ways. First, from a legitimation point of view, it is expected to meet with the greatest support. Secondly, 'just' refers to a kind of decision in which all relevant information has been adequately considered by the participants involved. In criticizing this approach, opponents have attended to what has been called the *'participation paradox'* (Seley, 1983): a just decision making procedure (e.g., a procedure with large involvement of affected groups and individuals) may not automatically lead to a just outcome, for some groups are more powerful than others. Participants confronted with lack of resources may therefore refrain from using their participation rights.

2.3 The market approach

The market approach stands quite differently, though not completely apart from the technical and public participation approaches. It refers to the common notion that people - despite the fact that they agree a site must be found somewhere - do not accept it in their own neighbourhood, because they are selfish. Some have called this phenomenon the NIMBY (Not In *My* Back Yard) syndrome (Popper, 1981). Siting a hazardous waste facility brings benefits and costs both to the local community near the site and to the national community as a whole. But the per capita costs for members of the local community are much higher than their per capita benefits. At the same time the per capita costs for local residents largely exceed the per capita benefits for people not directly involved. This has some important consequences.

The locals, confronted with an unproportionate burden, are prepared to resist the facility 'to the finish' (O'Hare et al., 1983), while no one outside the local group will be prepared to invest the same energy to realize it. Therefore, the locals have a good chance to 'win', which means that the siting of a hazardous waste facility becomes very hard if not impossible. From the market approach point of view public opposition occurs in large part because of the disproportionate cost-benefit-ratio local residents are confronted with. Equity should be provided by negotiating a 'fair price' for siting a hazardous facility. Public opposition could be overcome by offering the locals compensation and even reward. Then, benefits would outweigh costs also for the local population, so they could more easily accept the facility. Thus the market approach primarily considers the siting problem as a social dilemma (cf. Wilke, this volume).

In practice the market approach marches well together with the technical approach. Yet, there is one underlying principle that may lead to a rather deviant strategy: the market approach suggests people to be rational, though rational in the economic sense. This notion brings about the recommendation to negotiate with locals about a fair price for siting a facility. Some form of public participation is considered desirable and even necessary here, whereas the technical approach finds this to be the main cause of stagnation.

Suggesting free negotiations, however, supposes that both parties are willing to negotiate. But technical uncertainties could pose serious problems. It may be very difficult at least, to establish an adequate impact assesment and, in effect, a reasonable price. But what if local residents do not consider environmental and health impacts compensative at all? In that case, offers to compensate may even encourage local suspicion and aggravate conflict. Local authorities may be accused of bribery, and not always without reason.

Here we meet, what we refer to as the *market approach paradox*. Many people consider protection of health and safety to be one of the primary tasks of the welfare state. Free negotiations between officials and local authorities or citizens, however, pre-suppose the government either to withdraw or to behave itself as a private entrepreneur in a free market. Such behavior might well be considered illegal, if not in a judicial sense, then at least morally. In order to avoid such accusations, governments may regulate negotiations. But in effect the freedom of exchange will be hampered and so will the market approach itself.

2.4 The distributive justice approach

This approach, like the market approach, exploits the fact that siting a hazardous waste repository leads to an unequal distribution of gains and burdens. The fact that siting hazardous waste inevitably leads to disturbing equity is considered to be the main cause for public resistance. However, equity refers not only to the mutually exchangeable values referred to in the market approach, but also to health and safety of populations and generations: aspects that may not be considered negotiable by both parties.

The distributive justice approach admits that equity disturbance may not be compensated because it recognizes that equity could not be restored at all. Therefore, the utilitarian principle of justice which characterizes the market approach is rejected. Proponents of the distributive justice approach depart from Rawlsian equity principles (Kasperson, 1983, 1985). The most important derivation of these is that risks have to be avoided, and, if this is not possible, be spread over the whole population including those (but not exclusively those) who would benefit from a risky activity.

With this argument proponents of the distributive justice approach shift our attention from the public to the state. Not the market, but the elected government is considered as the very agent responsible for an equitable outcome of siting procedures. The state, because of its responsibilities for the protection of the public good, is the only agent capable of performing a long-term planning process assessing all impacts of risky activities, now and in the future, and providing measures in order to minimize and fairly distribute those impacts.

Sceptics of this approach put forward questions like: Who could impose equity principles on governmental agencies that represent competing interests? If informed consent is to be achieved, who will be responsible for the judgement of information and its transfer to the public? The distributive justice approach, they argue, imposes upon the state a task it may not be able to handle by its very nature as a democratic state. In stressing distributive justice in the Rawlsian sense, i.e., as a so-called end-state principle, this approach neglects crucial procedural matters. By neglecting the procedural problems of public participation, the distributive justice approach may well turn into the technical approach which also stresses "objective", "scientific" judgments and overlooks political controversy.

2.5 Contradictory approaches

So far we have explored four approaches all dealing with the siting problem from a public policy point of view. A first, brief comparison of these approaches reveals that they are partly complementary. For instance, we can imagine that in a real siting case the basic policy principle would be that the general public is incapable of making judgements on the technical issues involved (technical approach), but that at the same time local residents might claim compensation for losses in property value (market approach). An other example is a situation, in which the national government attempts to map all potentially hazardous locations in the country by making use of public participation (a combination of elements from the distributive justice and the public participation approach). Also other combinations of approaches appear to be possible siting strategies for policy makers.

But at the same time we may conclude that two combinations of siting approaches will lead to major difficulties, because these approaches are in fact antipoles. As we saw, the technical approach departs from the assumption that only professionals are capable of making rational judgements on risk, while the general public by definition reacts emotionally. The public participation approach, however, departs from the assumption that the people involved are able to make rational judgements on technical matters like risk, that they could also provide relevant information and that their judgement has to contribute largely to the final decision. Therefore, we will not expect the technical and the public participation approach to coincide into a public policy strategy concerning the siting of a hazardous facility.

This also appears to be the case with the market and the distributive justice approaches. Both, however, deal with a problem of justice. But both put forward an other solution for what in fact is not the same problem. The distributive justice approach aims for an equitable distribution of risks and benefits over the entire population. The market approach, however, focusses on the local community only. From the local perspective, there may be a situation of *injustice*, which from the national point of view may well be considered as just.

Others have referred to this contradiction between a distributive justice and a market approach as the contradiction between 'macro-justice' and 'micro-justice' (Brickman, Folger et al., 1981). What is good for the community as a whole is not necessarily good for every individual alike. In present-day societies the contradiction between macro- and micro-justice largely concerns the contradiction between state (national) regulation on the one hand and private (market) regulation on the other.

3. THE PUBLIC ACCEPTANCE OF HAZARDOUS FACILITIES

In many countries efforts to deal with hazardous siting decisions are of recent date. As a consequence relatively few studies are available which offer a thorough psychological analysis of siting issues and consequent policy recommendations.

A number of conclusions from psychological research, however, is of crucial importance for our analysis. Most psychological models describe the acceptability of risky activities as a trade-off between costs (including risks) and benefits and not solely as an evaluation of risks. This is true for the attitude models but also for the more cognitively oriented decision models (Otway, Maurer and Thomas, 1978; Vlek and Stallen, 1980; Derby and Keeney, 1981). Attitudes are described as a function of the subjective probability of expected consequences and the evaluation of these consequences. Perceived risks and expected benefits are interpreted as hardly separable negative and positive characteristics of a technology.

Most studies on large-scale technologies show for most respondents a dominant influence on attitudes of beliefs about threats to health, safety and the environment. With respect to nuclear waste the radiation risk is described as unknown, invisible and dangerous for health and environment, in the short as well as in the long run, and also with reference to genetic effects (Earle, 1981; Baillie, Brown and

Henderson, 1984). These largely intangible effects are frightening for many people. Economic consequences like economic growth, employment and energy tariffs are found to be included in the trade-off but nevertheless these seem to be of secondary importance (Midden, 1986).

Differences between supporters and opponents of nuclear power can be explained as differences in expectations of risks and socio-economic benefits. Their evaluation of consequences is very similar though. Supporters show a stronger tendency to make a trade-off between risks and benefits, whereas opponents' attitudes are mainly determined by (worries about) expected risks. Due to their uneven exposure to risks and benefits, the involvement of supporters is generally lower than that of opponents.

Which siting policy would result from this pre-dominance of safety and environmental risk perception? The relatively low perceived importance of the expected benefits of a facility leads to the expectation that people will not be very sensitive for incentives to compensate their exposure to hazardous waste facilities. As shown in a number of studies, proposals to compensate are not effective and may even be interpreted as bribery (Portney, 1985; Carnes and Copehaver, 1983). The idea of financial compensation underestimates the magnitude of people's risk assessment and its role in the overall risk-benefit trade-off.

Do these conclusions imply that compensation will never work as a strategy? The answer is probably negative. It is more likely that resistance to compensation only occurs if people really feel threatened. In siting procedures there will always be some local citizens who are not opposed to the use of the relevant technology. Usually these supporters find the risks tolerable and they strongly stress the economic benefits. From their viewpoint local interests could be served by counteracting siting decisions initiated on a central level. Socio-economic benefits could be optimized by negotiation. We suggest that compensation proposals are only adequate in dealing with socio-economic demands (Midden, Daamen and Verplanken, 1986).

As risk perception is the real issue for many residents who may become neighbours of a hazardous facility, some findings of risk perception research, which are especially important to understand people's reactions to risk bearing technologies, should be emphasized.

A crucial finding in risk perception research is the low correlation found between statistics of average yearly fatality rates and subjective judgements of seriousness of risk. One consequence might be that experts can not convince lay-people just by referring to their statistics. Obviously lay judgements are based on other factors (Slovic, Fischhoff and Lichtenstein, 1979).

A number of studies reveal that people's estimates of probabilities of the occurrence of serious accidents are much too high. On the whole it seems that low probabilities are overestimated and high ones are underestimated (see for a review Daamen, Verplanken and Midden, 1986). For experts, in comparison to lay-people, probability estimates are better predictors of attitudes towards the use of uranium. For the lay people feelings of insecurity are of great importance to their attitudes (Vlek and Stallen, 1981; Midden, Daamen and Verplanken, 1984).

A few comments need to be made about these results. First, in using "true frequencies" as a standard of accuracy one should realize that there may also be inaccurate frequencies. Second, some hazards are so new that no reliable statistical data are available and therefore the estimates of fatality frequencies are based on fault-tree analyses or comparable expert techniques. With such expert judgements certain error margins must be taken into account. Finally, we mention the problems with recorded frequencies as formulated by Green and Brown (1981): "The past does not define the future". A rational person will not mindlessly extrapolate the past into the future: risks may change.

A number of studies have revealed that catastrophic potential of risky activities is an important characteristic (e.g., Fischhoff, Slovic, Lichtenstein, Read and Combs, 1978; Renn, 1981; Vlek and Stallen, 1981). In other words, a large number of fatalities in a short period of time is judged as more serious than the same number spread out over a longer period of time. The resistance to nuclear power can partly be explained by its relatively great potential for disasters.

A second important factor underlying perceived risk is the expected controllability of consequences if something should happen. People tend to accept risky technologies more easily if they have a feeling of controllability. This expected controllability may be interpreted as personal control and/or as control delegated to experts (Midden et al., 1986). Important aspects of the expected controllability of radio-active waste are people's unfamiliarity with potential negative consequences and the invisibility of radio-active threats. Technological disasters may evoke, even more so than natural disasters, strong feelings of (actual) loss of control. After-effects of technological catastrophes may therefore be stronger than those of natural disasters.

Public participation in an open-information climate can play an important role in diminishing public distrust and feelings of uncontrollability. Witzig, Bord and Vincenti (1986) showed that residents of a planned site for low-level waste preferred power sharing options which increased their control, above incentives and compensation. The dilemma of the government was summarized by the authors as: "You can go through the public or over the public". This example also underlines a weakness of the market approach that we already referred to: worried citizens will not easily be prepared to accept serious risks, because financial compensation is offered. Besides it is hardly possible to express perceived risks, with their qualitative characteristics, into financial units.

Distrust is also a key issue in the distributive justice approach. The public will not accept judgements about justice and consequent procedures from a government which does not possess credibility among the public. Lack of credibility will lead to enhanced needs for participation. The public wants to be protected, but not by distrusted agencies.

4. SITING STRATEGIES

In the previous section we suggested that some approaches, e.g., the distributive justice and the public participation approach, would appear to be most successful in siting a hazardous technology, if the public perception of risk is taken into account. But, as will be shown below, the public perception of risk, though of crucial importance, is not the only factor that determines the success or failure of a particular siting strategy.

Other factors, concerning the aspect of *legitimacy*, will also appear to be important. These factors may be strong enough to affect the public perception of risk. To exploit these factors we have to look more closely at the rather complicated relation between public policy and public reactions. Therefore, in this section we present five siting cases, all being examples of combining different siting policy theories into a policy strategy, while also the public perception of risk may differ from case to case. In our presentation we focus on the two major contradictions between the policy approaches as discusssed in Section 2.5. We start with the problem of public participation, then we will observe the problem of justice.

4.1. The problem of public participation

Many siting cases include the technical approach, which apparently responds well to public officials' beliefs. Our first example reveals the experience in a rural county in the state of Virginia with a facility for household wastes (Jubak, 1982). Gradually the owner of the site came to use it for chemical wastes as well and some firms laid down proposals to found a chemical waste facility. It is important to note that initially the local inhabitants felt no objections at all. This rapidly changed as soon as it became clear that the County Board of Supervisors intended to settle the case in secret. Instead of organizing a public hearing for all local residents the Board got in contact with only a few of the nearest inhabitants. When this became public, apathy turned into resistance and the residents organized themselves into a local safety committee. The County Board changed its stand and began to support the local community. This largely contributed to the final result of the siting procedure. The local community successfully resisted the disposal of chemical wastes.

The second example (Brower Boyle, 1982) looks quite similar. Here, a big chemical firm tried to realize a site near Baltimore (Maryland). In the beginning the firm met with modest support from the local and regional officials. The firm wanted to anticipate negative reactions from the public. Therefore it organized information meetings, where it encouraged questions from the public to prove its credibility. But its proposal was rejected by the County Zoning Commission and at state level a law was proposed containing a prohibition to site a chemical waste facility at less than 1500 feet from a residential area. The firm lobbied the state governor to veto that law which would make the further siting procedure impossible. In addition, at the public hearings the firm offered a community park at the place of the landfills. But this offer only strengthened public resistance. People got the feeling of being bribed. Ultimately the firm decided to look for an other site.

In both cases the technical approach appears to have been dominant. In the Virginia case regional authorities intended to forego public opposition, trying to settle the case in secret. This proved to be an erroneous judgement. For the public became suspicious just because of that move. In the Maryland case the technical approach manifested itself in a somewhat deviant manner. The chemical firm apparently recognized it to be unwise to shun contact with local inhabitants. But it did not offer the opportunity for real public participation. Its intentions became manifest when the firm, with success, lobbied the state governor to veto a law that would have prevented all further endeavors at that site. In both cases we also glimpse at the market approach. The Virginia case shows a (secret) contact of county officials with the nearest locals, which might be revealing an intention to negotiate. The Maryland case shows the firm's intention to compensate for the negative consequences of the facility by offering a community park after the site had been completely filled. This, however, encouraged public suspicion.

In Virginia as well as in Maryland endeavors to site a chemical waste facility failed by using a policy strategy including the technical and the market aproach. The strategies failed, first of all because they did not take the public perception of risk into account. But, as the Virginia case showed, the public perception of risk became manifest at the moment residents became suspicious of the intentions of county officials. Apart from their risk perception, people apparently want to be taken seriously by governmental officials.

Our third example concerns a case where public participation was favoured instead of avoided. Sweden (Abrams, 1979; Paige and Owens, 1983) is still the only industrial country that has decided how to manage its radioactive waste on both the short and the long term. The decision became necessary, when in 1977 parliament passed a stipulation law holding that a continuation of the use of nuclear energy demanded an "absolutely" safe solution for the problem of radioactive waste. The joint electricity companies founded an organization for nuclear fuel safety, K.B.S., which in the same year published a proposal for long-term waste storage in rock formations.

Swedish government started a participation procedure that included a request for advice to many organizations within Sweden and other countries. The Swedish Energy Commission decided to use the method of 'Scientific Mediation'. This method is intended to clarify technical and scientific divergencies of opinion for politicians and the public, in order to settle the dispute politically in the most prudent way. Two scientists, a supporter and an opponent of the proposal, wrote down their personal opinion, the opponent being an American (the Swedish government did not expect to find competent opponents within Sweden itself). The result was a common paper where the differences of opinion including uncertainties were expressed in an open and clear way.

In the autumn of 1978 the Swedish government decided on a compromise and asked for further geological investigation. At the same time the left-wing and farmers' parties called for a referendum on the further use of nuclear energy which forced the government to resign. The majority of Swedish citizens in the referendum chose for the most moderate alternative that, still, called for a maximum capacity of twelve reactors until 2010 when other fuels would have to displace nuclear energy. At the same time a site for long-term storage of waste had already been chosen.

The Swedish policy strategy was characterized by an open information gathering process and an intensive political debate. One of the arguments put forward concerned the implications for future generations of the waste problem. This makes us conclude that the distributive justice approach is also present. We think that the Swedish case is the clearest example of a policy strategy combining the participation and the distributive justice approaches.

Are we allowed to conclude now that, because people want to be taken seriously, a siting strategy that consists of the public participation approach and the distributive justice approach is appropriate in dealing with the siting problem? The following examples will show that there are still some complications to deal with.

4.2 The problem of justice

Our fourth example shows that there are limits to public participation. People not only want to be taken seriously, they also want to be protected. This may well be illustrated by the failure of the Hazardous Waste Siting Facility Act (1980) of the state of Massachusetts (Bacow and Milkey, 1982). The innovative element in this act was that the State would provide local communities with financial means and expertise so that they could independently negotiate a site with a developer. The strategy is a combination of the public participation and the market approaches. Government withdraws in order to play a limited - though not unimportant - role in the background of the scene, regulating 'free' negotiations.

The contradiction appears to be felt by local citizens: negotiations can never be really free if regulations remain. At the same time, regulations are felt necessary because of the need for protection of the environment and public health. Thus, the government gives the impression that it acknowledges the right of public participation, but at the same time ignores its own obligations. This means that public perception of risk is acknowledged, but the protection of health and the environment has largely become the task of the local community and the private firm.

This 'market approach paradox' (cf. Section 2.3) occurs when a market approach strategy is implemented while at the same time distributive justice is an important part of current administrative practice. Under this condition people's resistance may be aggravated instead of softened, as was originally intended. People may feel that the government ignores its responsibility towards public health and the environment. This issue concerns the public perception of legitimate decision making and has not yet received serious attention in social scientific research.

Thus, public participation may not be extended so far that governmental responsibility for the protection of health and the environment comes into question. The protection people demand appears to be not only dependent on their perception of risk, but also on the normal administrative practices concerning environmental problems.

Our last case illustrates this point very well: what people consider 'just' or 'unjust' largely depends on the political system of which they are part. This case considers siting a nuclear power plant in the south of France, the Department of Midi-Pyrenées (Fagnani and Moatti, 1982). France appears to be the country where the technical approach manifests itself in a rather pure form. Highly centralized, and public participation procedures compared with other industrial democracies being only in embryo, the siting of nuclear power stations is not effectively opposed.

Nevertheless, until 1981 inhabitants, with support of the socialist-center majority in the Departmental council had been able to resist the power plant. Shortly after the socialists came into power, the Council changed its mind. It accepted the plant on condition of written guarantees that it would benefit the region. In February 1982 a delegation of Electricité de France (E.D.F.) and the socialist chairman of the Departmental Council signed an agreement. It contained several measures to prevent environmental damage concerning the fish in the Garonne river, compensation orders to an amount of 1.2 billion francs and the hiring of a fixed percentage of local workers for constructing the plant as well as during its operation period.

The agreement also contained a secret section that was published by "Le Monde" two months later. E.D.F. provided the Department over Fr. 6 billion in payments each year during the operational stage of the plant. The agreement was brought about easily because the socialist chairman of the Departmental Council became chairman of the environmental finance commission in national Parliament at the same time. All this provoked some political reactions - especially the secret part - merely because the price was considered too high. In reaction to the generosity of E.D.F. the minister of Energy Affairs stated this case had to be judged as an exception due to the absence of adequate procedures.

In this case we observe the technical approach, which dominates French environmental and energy policy. The French strategy also contains an element of distributive justice, for the state policy is the result of an overall planning process, though centralized and neither open for public participation nor for large parliamentary involvement. During the seventies, the government started to compensate regions that accepted nuclear power plants. People were provided with a 15% reduction on their electricity bill. This might be considered to be an element of the market approach, but it must be noticed that freedom to negotiate - a significant feature of the market approach - was absent.

The socialist government expressed the intention to move towards more decentralization and open administrative procedures. By Spring of 1982 this intention had not yet led to new legislation. Within this context the market approach appeared. The freedom to negotiate was limited to the regional government (e.g., the socialist and radical party leaders); the local residents were not asked to express their views at all. The success of this strategy, therefore, was due to the political and administrative context that envisioned the technical as well as some elements of the market approach.

5. CONCLUSIONS

Now we can reveal some factors that may determine the effectiveness of a government siting strategy. The first of these factors, without any doubt, is the *public perception of risk*. People will tend to resist a facility siting only if they are anxious about possible negative impacts on their living conditions. It can be derived from the above cases that the perception of risk for health and the environment may especially contribute to heavy resistance. Perceived economic losses may stimulate negotiations about compensation for the local community. But beside the public perception of risk some factors at the institutional level appear to be of crucial importance as well.

It was demonstrated that two combinations of policy theories can not combine in a real policy strategy, because the underlying approaches are contradictory. The technical and the public participation approaches both deal with the problem of public participation in an opposite way, while the market and the distributive justice approaches reveal an opposite perspective on the problem of equity, stressing micro- and macro-justice, respectively.

A policy strategy combining the public participation and the distributive justice approaches, may look most attractive from a public perception point of view. As we noticed, these approaches take the public perception of health and environmental risk most seriously. But governments, of course, not only draw upon public perception of risk in the development and implementation of public policy. They also consider national and international laws and regulations, political customs and the necessity for coherence of the different public policy fields. The above case studies show that this *administrative context* is the second factor, relevant for the success or failure of a siting strategy.

The third factor of importance for the success or failure of a siting strategy is *trust in government*. Distrust and suspicion occur when public officials try to settle the case in secret (cf. the Virginia case). But, on a more general level, distrust may occur as large parts of the population and the government are no longer talking about the same subject. This is not only a matter of public perception, but also of governmental strategy. In the Massachusetts case, the government, by introducing the market approach, violated not only the expectations of the general public about govermental handling of environmental problems, but also its own administrative practice.

Thus we conclude that three factors appear to determine the success or failure of a governmental siting strategy: (1) the public perception of risk, (2) the administrative context of public policy making, and (3) trust in the government. Given this framework we can not indicate one single strategy as being the most effective one.

Our observations lead us to the conclusion that nations are confronted with two main dilemmas: governmental agencies may (1) widen or narrow the opportunities for public participation, and (2) increase or decrease their own role in the siting process. Figure 1 shows how these dilemmas are related to the different siting strategies.

Figure 1: Four types of policy strategies.

```
                        GOVERNMENT INTERFERENCE
                 LARGE                        SMALL

              ┌─────────────────────┬─────────────────────┐
              │          A          │          B          │
              │                     │                     │
              │    DISTRIBUTIVE     │       MARKET        │
              │  JUSTICE APPROACH   │     APPROACH        │
  LARGE       │                     │                     │
              │    PARTICIPATION    │    PARTICIPATION    │
              │      APPROACH       │      APPROACH       │
  PUBLIC      │                     │                     │
PARTICIPATION ├─────────────────────┼─────────────────────┤
              │          C          │          D          │
              │                     │                     │
              │    DISTRIBUTIVE     │       MARKET        │
              │  JUSTICE APPROACH   │     APPROACH        │
  SMALL       │                     │                     │
              │     TECHNICAL       │     TECHNICAL       │
              │     APPROACH        │     APPROACH        │
              └─────────────────────┴─────────────────────┘
```

We expect these dilemmas to be present in any political system trying to deal with a problem of siting a hazardous technology. But we can not indicate a general direction in which they may be resolved. Among most industrial nations, however, the Swedish strategy proved to be most successful. It implied increasing public participation as well as an active role for political institutions. The choice for such a strategy may be difficult. Governments may rather choose to switch to a market-approach strategy, if they consider the costs of public participation to be too high. Lack of time and the risk of enduring conflict may be arguments to follow an other path. But choosing an other strategy has its consequences as well. From the values, expressed by 'democracy', effectiveness is just one. Therefore, siting strategies that are widely considered to be legitimate and just, may appear to be the most effective in the long run.

REFERENCES

Abrams, N.E. (1979). Nuclear politics in Sweden. *Environment, 22(4)*, 6-40.

Bacow, L.S. and Milkey, J.R. (1982). Overcoming local opposition to hazardous waste facilities: The Massachusetts approach. *Harvard Environmental Law Review, 6*, 265-305.

Baillie, A., Brown, J. and Henderson, J. (1984). *Perception of nuclear power and the management of information*. Department of Psychology, University of Surrey, Guildford.

Brickman, P., Folger, R. et al. (1981). Microjustice and macrojustice. In: M. Lerner and S. Lerner (Eds.), *The Justice Motive in Social Behavior*. Boston: Allyn Bacon.

Brower Boyle, S. (1982). *An analysis of siting new hazardous waste management facilities through a compensation and incentives approach*. Cornell University, Ithaca, New York.

Carnes, S.A., Copehaver, E.D. et al. (1983). Incentives and nuclear waste siting: Prospects and constraints. *Energy Systems and Policy, 7(4)*, 323-351.

Daamen, D.D.L., Verplanken, B., and Midden, C.J.H. (1986). Accuracy and consistency of lay estimates of annual fatality rates. In: B. Brehmer, H. Jungermann, P. Lourens, and G. Sevón (Eds.), *New directions in research on decision making* (pp. 231-243). Amsterdam: North Holland.

Derby, S.L. and Keeney, R.L. (1981). Risk analysis: understanding "how safe is safe enough". *Risk Analysis, 1*, 217-224.

Earle, T.C. (1981). *Public perceptions of industrial risks: The context of public attitudes towards radioactive waste*. Seattle, Washington:

Battelle, Human Affairs Research Center.

Edwards, W. and Von Winterfeldt, D. (1985). Public disputes about risky activities: stakeholders and arenas. In: V.T. Covello, J.L. Mumpower, P.J. Stallen and V.R.R. Uppuluri (Eds.), *Environmental impact assessment, technology assessment and risk analysis* (pp. 877-912). Heidelberg/New York/Tokyo: Springer.

Fagnani, J. and Moatti, J.P. (1982). *France, The socialist government's energy policy and the decline of the anti-nuclear movement.* Working conference ECPR, Berlin.

Fischhoff, B., Slovic, P., Lichtenstein, S., Read, S. and Combs, B. (1978). How safe is safe enough? A psychometric study of attitudes towards technological risks and benefits. *Policy Sciences, 8,* 127-152.

Green, C.H. and Brown, J.(1981). *The perception and acceptability of risk.* Dundee (U.K.): Duncan of Jordanstone School of Architecture.

Jubak, J. (1982). Struggle over siting hazardous waste disposal. *Environmental Action, Feb.,* 14-17.

Kasperson, R.E., (Ed.). (1983). *Equity issues in radioactive waste management.* Cambridge (Mass.): Oelgeschlager, Gunn and Hain.

Kasperson, R.E. (1985). *Rethinking the siting of hazardous waste facilities.* Conference Institute for Applied Systems Analysis, Vienna, Austria, July 2-5, 1985. Clark University, Worcester, U.S.A.

Leeuw, F.L. (1986). Reconstructing and evaluating policy theories as an instrument of social impact assessment. The case of future population policy in the Netherlands. In: H.A. Becker and A. Porter (Eds.), *Impact Assessment Today* (pp. 103-122). Utrecht: Van Arkel.

Midden, C.J.H. (1986). *Individu en grootschalige technologie, een vergelijkend attitude-onderzoek naar de opwekking van elektriciteit met kolen en uraan.* Dissertation, Leiden.

Midden, C.J.H., Daamen, D.D.L. and Verplanken, B. (1984). *De beleving van energierisico's.* Leidschendam (Neth.): Ministerie van Volkshuisvesting, Ruimtelijke Ordening en Milieubeheer; Publicatiereeks Milieubeheer, 3.

Midden, C.J.H., Daamen, D.D.L. and Verplanken, B. (1986). *Public reactions to large scale energy technologies.* Paper presented at the 13th Congress of the World Energy Conference, Cannes, France, October 5-11, 1986.

O'Hare, M., Bacow, L. and Sanderson, D. (1983). *Facility siting and public opposition.* New York: Van Nostrand Reinholt.

Otway, H.J., Maurer, D. and Thomas, K. (1978). Nuclear power: the question of public acceptance. *Futures, 10*, 109-118.

Paige, H.W. and Owens, J.E. (1983). *Assessment of national systems for obtaining local siting acceptance of nuclear waste management facilities*. Final report. Vol.1. Background on political structure and formal system for approving waste management siting decisions. Vol.2. Summary of principal new developments relating to the siting of waste management facilities (1.1.82-4.1.83). Washington D.C.

Popper, F. (1981). 'Siting LULUS'. *Planning, 47(4)*, 12-15.

Portney, K.E. (1985). The potential of the theory of compensation for mitigating public opposition to hazardous waste treatment facility siting: some evidence for five Masschusetts' Communities. *Policy Studies Journal, 14*, 81-89.

Renn, O. (1981). *Man, technology and risk*. Report of the Nuclear Research Centre (Jül-Spez-115). Jülich, Federal Republic of Germany.

Seley, J.F. (1983). *The politics of public-facility planning*. D.C. Heath and Company, Lexington, Mass./Toronto.

Slovic, P., Fischhoff, B. and Lichtenstein, S. (1979). Rating the risks. *Environment, 21(3)*, 14-39.

Susskind, L. and Elliott, M. (1983). *Paternalism, Conflict and Coproduction, Learning from Citizen Action and Citizen Participation in Western Europe*. New York and London: Plenum Press.

Van der Pligt, J. (this volume). Nuclear waste: public perception and siting policy.

Van de Vall, M. and Ulrich, H.J. (1986). Trends in data-based sociological practice. Towards a professional paradigm? *Knowledge: Creation, Diffusion, Utilization, 8(1)*, 167-184.

Vlek, Ch. and Stallen, P.J. (1980). Rational and personal aspects of risk. *Acta Psychologica, 45*, 273-300.

Vlek, Ch. and Stallen, P.J. (1981). Judging risks and benefits in the small and in the large. *Organizational Behavior and Human Performance, 28*, 235-271.

Wilke, H.A.M. (this volume). Promoting personal decisions supporting the achievement of risky public goods.

Witzig, W.F., Bord, R.J. and Vincenti, J.R. (1986). Public perception of low-level waste technologies: demands in research and public education programmes. *Proceedings of the Symposium on Waste Management at Tucson* (pp. 69-74), Arizona, March 2-6.

Zijlstra, G.J. (1982). *The policy structure of the Dutch nuclear energy sector*. Dissertation, University of Amsterdam.

Zinberg, D.S. and Deese, D. (1980). *Radioactive waste management, a comparative study of national decision making processes*. Final report for period September 15, 1978 - December 31, 1979. Harvard University, Cambridge, Mass.

NOTES

1. Our arguments are largely based upon the findings of a study entitled: Het kiezen van lokaties voor gevaarlijk (radioaktief) afval (Siting decisions for radioactive waste), financed and published by the Department of Housing, Physical Planning and Environmental Affairs, The Hague, The Netherlands. We acknowledge the contribution of Pieter Jan Stallen who co-authored that study.

PLANNING AND PROCEDURAL ASPECTS OF THE WINDFARM PROJECT OF THE DUTCH ELECTRICITY GENERATING BOARD

FRITS LUBBERS

Division of Physical Planning and Environmental Affairs
Dutch Electricity Generating Board (Sep), Arnhem

1. INTRODUCTION

This chapter gives the full story of the preparatory and executive decision making processes concerning a novel and noncontroversial electric-power project: the siting of an experimental wind turbine park. The author's contribution proper starts at Section 1.1. By way of an editorial preface we (Ch.V. and G.Cv.) would like to note beforehand that roughly two phases appeared to be necessary which seem to have been conducted to the satisfaction of the several parties involved (see Section 4).

In the first - planning - phase, described in Section 2, design and evaluation criteria for determining a 'best' site were developed. Here, technical-scientific, environmental and economic aspects turned out to be essential for a thorough multi-criteria (or multi-attribute utility) analysis which concluded phase 1.

In phase 2, described in Section 3, the selected 'best' site, as a government proposal, was subjected to various formal political-administrative procedures required for obtaining the necessary licences. This brought the proposal into the arena of evaluation committees, provincial and municipal authorities, and groups of citizens at public hearings.

The two-phase process of judgment and decision making nicely illustrates one way of tackling the *individual judgment problem and the social interaction and aggregation problem* as distinguished in chapter 1, for an actual technology siting problem in the lively political context of the energy debate. In Sections 1.1 and 1.2 the paper starts with a description of the general background and objectives of the windfarm project.

Ch. Vlek and G. Cvetkovich (eds.), Social Decision Methodology for Technological Projects, 195–215.
© *1989 by Kluwer Academic Publishers.*

1.1 General Background of the Windfarm Project

In August 1981 the Dutch Government set up the interdepartmental Commission for Wind Energy and Storage. This Commission was to prepare the Government's position on the basis of the study "Wind Energy and Hydro-power" (Begeleidingscommissie Voorstudie, 1981), which was published in May 1981. The Government's position in "Wind Energy and Storage" was presented by the Minister of Economic Affairs on July 7, 1982 (Tweede Kamer, 1982).

In principle the Government took a positive attitude towards the use of wind energy for the production of electricity in the Netherlands. Diversification, fuel savings, environment, industrial employment and innovation were mentioned as important arguments for this standpoint. In this context the Government held the opinion that a windfarm project for demonstration purposes should be realized as soon as possible. This demonstration plant would be an essential first step in the development of large-scale wind energy facilities in the Netherlands.

Furthermore, the Government held the opinion that the Dutch Electricity Generating Board (Sep) or a company related to Sep should direct the realization and operation of this demonstration project. During the preparation of the Government's viewpoint the Sep management was contacted. These contacts revealed that the ideas of Sep concerning the development of wind energy were largely in line with those held by the Government.

In June 1982 the Sep Board of Commissioners and the meeting of shareholders agreed to the proposal by the Sep management for realizing a windfarm with a capacity of about 10 MegaWatt (MW), (N.V. SEP, 1983). This consent was given on the condition that between June 1982 and December 1983 a better understanding would be obtained of the techno-economic limiting conditions which the project had to meet, and the chances of obtaining the required licences.

As a result of the above-mentioned decisions, the Ministry of Economic Affairs and Sep developed a joint project to build a windfarm with a total capacity of about 10 MW.

1.2 Objectives of the Windfarm Project

The main objective of the windfarm project is to come to a well-founded decision, on the basis of experience gained in building and operating the windfarm, concerning the feasibility of wind energy for electric power generation in the Netherlands.

This overall goal includes the following specifications:
- To study the effects of park and line configurations in such a way that the results may be considered representative of large windfarms and of configurations of lines with large wind turbines.
- To prepare the way for further developments and the extension of Dutch industrial wind turbine activities.
- To gain experience with planning aspects and procedures, environmental aspects and social acceptance.
- To gain a better view of the cost of medium-sized and large wind turbines in mass production.

2. ORGANIZATIONAL AND PLANNING ASPECTS OF THE WINDFARM PROJECT

2.1 Organization

The preliminary work for the windfarm project was supervised by a policy committee. Decisions concerning activities related to the windfarm project were taken by this committee. The Government and Sep each appointed half of the members of this committee, which in its turn set up a number of advisory committees.

One of these committees, the Advisory Committee for Social and Environmental Aspects (AMMA), was set up to advise the policy committee, especially on matters related to gaining experience in planning aspects and procedures, environmental aspects and social acceptance. AMMA consists of a number of independent experts. Their main task is to advise the policy committee on the above-mentioned aspects of the project. In the light of future possibilities of wind energy, judging and especially evaluating these aspects is highly important for electricity supply in the Netherlands.

2.2 Structure of Site Exploration for the Windfarm Project

2.2.1 Basic Assumptions

In making their conditional decisions the Sep Board of Commissioners stipulated that the exploration of possible sites for the windfarm project had to be nation-wide. Every possible site in the Netherlands was to be included in this exploration. This nation-wide exploration took place from June 1982 until December 1983.

The anticipated windfarm project consisted of about 24 horizontal-axis wind turbines with a capacity of between 500 and 300 kW each and a rotor diameter and axis height of between 25 and 35 m. This would require approximately 60 ha of space (600 x 1000 m).

In exploring the possible sites, two basic assumptions were formulated:
- A possible extension of the windfarm to about 250 ha (1000 x 2500 m) was taken into account, which ensures the necessary flexibility of the site chosen.
- The site use should be multifunctional. Should the use of wind energy prove feasible in the Netherlands in the future, then the windfarm must in effect be arranged in such a way that the same area may still be used for other (especially agricultural) purposes.

2.2.2 Structure of Site Exploration for the Windfarm Project

The exploration of possible windfarm sites consisted of three phases. In the first phase site requirements were drafted by Sep, which could be used to make an inventory of possible sites by the Provincial Electricity Generating Companies, in cooperation with the provincial Planning Boards. For this inventory the national screening analysis, which had been used for the Government's viewpoint in "Wind Energy and Storage" (Tweede Kamer, 1982) was used. The inventory eventually yielded approximately 60 potential sites.

In the second phase a rough-cut selection based on the site requirements was made. The approximately 60 potential sites were selected according to the following criteria:
- *General requirements* for windfarm sites:
  * there should be sufficient wind (the average annual wind speed should exceed 6.5 m/sec at an altitude of 40 m;
  * no special use should be made of the area (airports, sea dikes,

military shooting ranges, woods and water);
* sites should be outside of airport approach routes;
* sites should be outside of telecommunication connections.
- *Special requirements* resulting from the windfarm's being a demonstration project:
  * a surface of at least 1 x 2.5 km (including space for extension), with the greatest length perpendicular to the prevailing wind direction; furthermore, consistent ground roughness for 1 km in the direction of the prevailing wind;
  * no occasional buildings or houses within the 60 ha windfarm area;
  * no occasional buildings within 250 m around the 60 ha windfarm area;
  * no continuous buildings within 1000 m around the 60 ha windfarm area;
  * no building in areas for special use, such as scenic areas, potential national parks, (potential) nature reserves, open areas and buffer zones.

After this examination, in which the surface requirement proved to be the most prohibitive, 8 possible sites remained. These sites, which were all considered more or less suitable for a windfarm project, are listed in Table 1 below (Section 2.4.1).

In the third phase a detailed examination of this selection of eight suitable sites was made. The examination consisted of an assessment and evaluation of these eight sites on the basis of the preferential site characteristics, and finally resulted in a preference order for the eight sites, as well as in a choice of 3 sites, for which procedures were to be started. The detailed examination is discussed extensively in the next section.

2.3 Screening of the Eight Selected Suitable Sites

The preferential characteristics for the windfarm site formed the touchstone of the detailed examination. These characteristics can be divided into three categories: research characteristics, techno-economic characteristics, and hindrance characteristics. The evaluation of a site was determined by the degree in which the different requirements in all three categories were met.

2.3.1. Research Characteristics

In connection with the research characteristics three factors are important:
- *Ground roughness*: To insure optimum reliability of the results of the experiments with respect to windfarm, it is important that the ground in the surroundings of the windfarm, up to 1 km in the prevailing wind direction, consists of pastures or fields, preferably with a roughness grade of 3 to 4. The smaller the influence of external factors, such as roughness, on the experiments, the better and more reliably the investigations can be carried out.
- A site with possibilities for experimenting with *interaction between birds and wind turbines* is to be preferred to a certain degree: The investigation of this problem is important, as it has not yet been established whether the interaction between birds and wind turbines calls for a positive or a negative approach, i.e., a site with sufficient quantities of birds is to be preferred.
- An other important aspect for the feasibility of wind energy in electricity supply in the Netherlands is *the social perception and acceptance of the wind-energy option*. A site where its acquaintance can be made in a relatively easy and informal way is to be preferred. This means that sites located centrally in the Netherlands, for instance near motorways or densely populated areas, should have relative priority.

2.3.2 Techno-economic Characteristics

In connection with the techno-economic characteristics six factors are important:
- *The wind regime*: Because of the fact that the output of wind turbines increases by a third power with increasing wind speeds, the site with the highest wind supply has priority from an economic point of view.
- *Connection to the national power grid*: As the connection of the windfarm on to a switching and transformation substation is extremely expensive, a site which is close to an existing high-voltage station has relative priority.
- *Infrastructure of the windfarm*: In building and maintaining the windfarm, it is important to have a good infrastructure at and around the windfarm. A site with a good infrastructure has priority.
- *Soil structure*: The soil structure is relevant in view of the construction costs to be incurred, for instance for the foundations

of the wind turbines and the infrastructure. The site where these costs are lowest has relatively greater priority.

- *Possibilities of extension*: In view of flexibility and for economic reasons, a possible extension of the existing windfarm project in the area surrounding it is preferred on account of the capital costs already invested in infrastructure, connections, etc. From this point of view a site with optimum possibilities for extension is to be preferred.
- *Accessibility*: For builders, experimenters and visitors accessibility is an important economic aspect. Preference is therefore given to the site which is most easily and rapidly accessible, and at the lowest cost.

2.3.3. Hindrance Characteristics

With regard to the hindrance characteristics five factors are important:

- *Inconvenience to the surroundings*: A site where possible inconvenience to the neighbourhood, caused by noise and vibrations can be reduced to a minimum is preferable.
- *Agriculture*: A site where the rigid configuration of the windfarm fits best into the existing structure of lots and other landscape structures is preferred from the viewpoint of hindrance for agriculture. Affecting the structure of the lots/landscape could have serious consequences for the degree of mechanization in agriculture.
- *Aspects concerning the landscape*: Carrying out the windfarm project means that a new, eye-catching and modern element will be added to the Dutch landscape. The hindrance caused by the windfarm project is generally defined in terms of the degree in which certain essential characteristics of the landscape are affected. The following aspects are especially important in this hindrance-factor: visual urbanization, natural quality of the landscape, structure and scale of the landscape, openness of the landscape, appreciation of the landscape and cultural-historical value.
- *Damage and nuisance to birds*: The site which causes the least damage and nuisance to birds is preferred from this point of view. Special preference is given to sites outside of the sleeping- and fooddrift-routes.
- *Government policy*: Preference for a certain site is greatly dependent on the extent to which a site is in line with the Government's environmental and physical planning policy.

At the end of this review of site requirements and preferential site characteristics it may be said that the stands taken by communities and provinces during the granting of the necessary licences also played an important role. As already stated above, the general and special site requirements played an important role as prohibitive site criteria in the first general selection from the 60 sites listed. The preferential site characteristics also played a role in the first investigation. During the detailed examination, however, these characteristics played a much *larger* role. In this phase an assessment and evaluation of the eight sites was made on the basis of these preferential site characteristics.

2.4 Description of the Analytic Method: Multi-criteria Analysis

Multi-criteria analysis is generally used in environmental and physical planning if different options exist and there is a need to come to a partial or total priority ranking of these options (Keeney and Raiffa, 1976; Voogd, 1980; Zeleny, 1982). The multi-criteria analysis is also an answer to the restrictions of a cost-benefit analysis. Unlike a cost-benefit analysis, qualitative information can be compared in the multi-criteria analysis. This sort of analysis is therefore especially useful for the evaluations needed to determine the choice of a site. In view of the information which was available on the windfarm sites, the multi-criteria analysis is the most eligible method for a preference selection.

Qualitative information (ordinal data), however, can not in fact be used for arithmetical calculations. To be able to manipulate the data so as to arrange or eliminate options, it is necessary to make a few assumptions. One of these assumptions is to consider the ranking numbers of the ordinal scale as units on a ratio-scale. This makes it possible to use many quantitative methods, but it diminishes the ordinal character of the data. A refinement, which still maintains these assumptions, is to revalue the orders of priority, for instance through paired comparisons, thus arriving at data which can be treated quantitatively. This is done in the so-called ordinal concordance method (Roy, 1968). It may happen that a small variation in the evaluation of criteria results in a different order of priority of the alternatives. Therefore, it is always useful to work out the order of priority from different points of view. A certain viewpoint can be characterized by making one aspect extremely important.

There are techniques for coupling a matrix with assessments by site to the weights which can be given to the site-criteria. Because of the nature of the assessments, which are virtually all non-quantitative, the starting point consisted of ordinal assessments. Two types of methods were used:

a. *The expected value method with (ordinal) orders of priority*
   This method gives a general picture and should be used very carefully. The information obtained from the assessment matrix is quite "rough", which means that more subtle gradations in information are lost. Therefore, the expected value method should always be used in combination with other methods.

b. *Concordance methods (the ordinal Electre method; Roy, 1968)*
   This method or analysis deals more "carefully" with the information obtained from the assessment matrix. In this method, the different versions are compared with each other in couples. Two questions are raised in determining to what degree an option is better or worse than, or equal to, an other option. First, there is the question to what extent one option is better than an other according to a selected criterion. Second, there is the question of which characteristics are preferred in the evaluation of the various options.

In this site selection procedure the concordance method has been used. This has resulted in a preference matrix in which approximately thirty sensitivity analyses were carried out, each emphasizing a different priority, and resulting in a specific order of preference across options.

2.4.1 Assessment of Sites per Characteristic

The assessment of the sites took place on the basis of the preferential site characteristics. The ratings were made by Sep based on common judgements by various persons. Because differences between the assessments of some sites can not always be expressed quantitatively nor consistently, the ordinal method was used in this evaluation.

The desire to keep subjective assessment at an absolute minimum level and thus increase the objectivity of the examination, was an additional reason for choosing this ordinal method, so that a multi-criteria analysis could be used in the evaluation phase. On the basis of the preferences for certain sites, an order of priority of the eight remaining sites was established. All this is summarized in the assessment matrix of Table 1.

Table 1: The assessment matrix.

| | ALTERNATIVE SITES* | | | | | | | |
CHARACTERISTICS	A	B	C	D	E	F	G	H
HINDRANCE								
GOVERNMENT POLICY	2	2	1	3	4	1	1	2
BIRDS	5	5	2	4	4	1	1	3
ASPECTS REGARDING LANDSCAPE	2	4	1	3	3	3	3	3
AGRICULTURE	1	1	2	3	3	3	3	2
INCONVENIENCE TO SURROUNDINGS	4	3	3	1	2	3	3	5
TECHNO-ECONOMIC ASPECTS								
ACCESSIBILITY (BUILDERS, VISITORS, EXPERIMENTERS)	4	4	2	1	1	3	3	5
POSSIBILITIES FOR EXTENSION	3	3	4	1	3	2	6	5
SOIL STRUCTURE	3	3	2	3	1	2	2	1
INFRASTRUCTURE ON THE FARM	6	7	1	4	4	5	2	3
CONNECTION TO THE NATIONAL POWER GRID	4	4	3	4	5	5	2	1
WIND	1	1	2	2	2	1	1	1
RESEARCH								
SOCIAL PERCEPTION	3	3	1	1	1	2	2	4
BIRD RESEARCH	1	1	4	2	2	5	5	3
ROUGHNESS	1	2	3	2	2	2	2	2

* **A**=Sexbierum, **B**=Alchum, **C**=Noordoost Polder, **D**=Southern Flevoland, **E**=Southern
   Flevoland/Lake Eem, **F**=Wieringermeer 1, **G**=Wieringeremeer 2, **H**=Hoogland Polder

2.4.2 Sensitivity Analysis of the Site per Characteristic

The assessment matrix (Table 1) was the basis for a further evaluation
of the sites. This evaluation of sites carried out by Sep was necessary
to obtain further insights into the priority order of the sites. The
three categories of site characteristics (i.e., research-related,
techno-economic and hindrance characteristics, see Section 2.3) were
submitted to a sensitivity analysis by means of a multi-criteria
analysis. The aim of the sensitivity analysis was to check the
consistency of the different priority orders by emphasizing different
aspects.

In each sensitivity analysis the evaluation was changed in such a way
that one of the site characteristics or groups of site characteristics

was the most important. Thus, from various points of view a preferred order of priority of the sites reviewed was obtained. This analysis of the different options was very important, as the numerous qualitative factors precluded a direct evaluation.

Table 2: Review of the factors of DELPHI-weighting (the higher the mark, the more important the aspect).

	DELPHI-WEIGHTING
HINDRANCE ASPECTS	
GOVERNMENT POLICY	5
BIRDS	3
ASPECTS REGARDING LANDSCAPE	1
AGRICULTURE	6
INCONVENIENCE TO SURROUNDINGS	1
TECHNO-ECONOMIC ASPECTS	
ACCESSIBILITY (BUILDERS, VISITORS, EXPERIMENTERS)	1
POSSIBILITIES FOR EXTENSION	2
SOIL STRUCTURE	3
INFRASTRUCTURE ON THE FARM	1
CONNECTION TO THE NATIONAL POWER GRID	3
WIND	6
RESEARCH	
SOCIAL PERCEPTION	3
BIRD RESEARCH	2
ROUGHNESS	6

A "limited DELPHI-method" formed the basis for the evaluation (DELPHI-weighting). Six independent experts from different backgrounds made their own preference lists of the weighting factors with relevant supporting arguments. In the following session these preference lists and arguments were discussed in the group as a whole, and then each list was again made separately. This procedure was repeated several times, until a consensus concerning the preference list of weighting factors was finally arrived at. The results of the DELPHI-weighting are represented in Table 2. In this way the different aspects were calibrated in comparison with one another. Afterwards, this general importance weighting was adapted to yield the rank orderings of sites according to different viewpoints.

DELPHI-weighting may be considered as a basic weighting procedure.

By means of the multi-criteria analysis it is possible to give a certain characteristic a higher preference value than the others. Sep has investigated the following versions:
- The *"Wind-force 9"* version, in which the characteristics 'roughness' and 'wind' are important.
- The *"Bird-friendly"* version, in which 'hindrance to birds' is important.
- The *"Bird-free"* version, in which 'hindrance to birds' is less important.
- The *"Beets"* version, in which 'hindrance to agriculture' is important.
- The *"Confrontation"* version, in which 'better access to the windfarm' is important.
- The *"Chamberlain"* version, in which the importance of 'staying in line with Government policy' is accentuated.

The results of these analyses are presented in Table 3.

Table 3: Preference Matrix on the eight possible sites, A-H, resulting from the ordinal multi-criteria analysis.*

	VERSION						
	1	2	3	4	5	6	7
Highest	A	A	G	A	A	C	G
Preference	G	G	F	B	H	G	C
	H	H	C	H	C	F	F
	F	F	H	C	G	A	A
	C	C	A	G	B	D	H
	B	B	B	F	F	E	B
Lowest	D	D	D	D	D	H	D
Preference	E	E	E	E	E	B	E

* The symbols A to H and 1 to 7 have been used for the following variables:

A = Sexbierum	1 = general DELPHI-weighting
B = Alchum	2 = Wind-force 9 Version
C = Noordoostpolder	3 = Bird-friendly Version
D = Southern Flevoland	4 = Bird-free Version
E = Souther Flevoland/	5 = Beets Version
Lake Eem	6 = Confrontation Version
F = Wieringermeer 1	7 = Chamberlain Version
G = Wieringermeer 2	
H = Hooglandpolder	

In the multi-criteria analysis it is also possible to give a certain group of site characteristics a higher preference value. The following versions have been investigated by Sep:
- The *"Poltergeist"* version, in which the reduction of hindrance is important, and which yielded the 'Noordoostpolder' site as the most preferred option;
- The *"Money"* version, in which the techno-economic factors are important, which led to 'Wieringermeer 2' being the best option;
- The *"Research-is-illuminating"* version, in which the importance of the experimental and evaluation character of the windfarm project is accentuated, and coming out with 'Sexbierum'; and
- The *"Money-is-not-important"* version, in which the combined characteristics of research and hindrance are accentuated, and which also yielded 'Sexbierum' as the most attractive site.

2.5 Results

On the basis of the multi-criteria and sensitivity analyses the following conclusions can be drawn. The evaluation per preferential site characteristic can be divided into three parts, depending on the degree of aggregated preference for the sites evaluated:
a. sites with the highest preference: Sexbierum, Noordoostpolder and Wieringermeer 2;
b. sites with a low preference: Wieringermeer 1 and Hooglandpolder (Zeeland);
c. sites with the lowest preference: Alchum, Southern Flevoland, Southern Flevoland/Lake Eem.

On the basis of the eight versions presented, the difference in preference between Sexbierum, Noordoostpolder and Wieringermeer 2 turns out to be very small: Sexbierum is a slightly preferable site from the point of view of wind supply, the possibilities of experiments concerning bird hindrance, and agricultural interests (Table 3, columns 2, 4 and 5 respectively). Wieringermeer 2 is a preferable site from the point of view of Government policy (column 7) and reduction of bird hindrance (column 3). Noordoostpolder is a slightly preferable site because of its accessibility (column 6).

These conclusions are supported by the sensitivity analysis by group of characteristics in the Poltergeist, Money, Research-is-illuminating, and Money-is-not-important versions (see above).

On the basis of the results of this evaluation, it was decided to continue the procedures at the Sexbierum, Noordoostpolder and Wieringermeer 2 sites and, in view of the experimental character of the windfarm project, to express a preference for the site at Sexbierum since this site was the most preferable under the combined characteristics of research and hindrance reduction.

As this analysis was made with an assessment matrix containing many 'equivalent' equal assessments for different sites, there was the possibility that quite relevant information would be covered up because of this. To counteract the effect of indiscriminate assessment, a control analysis was performed on the results, by means of an assessment matrix in which all sites were given a different priority rank for each characteristic, even when the differences were felt to be extremely small.

A concordance analysis with the same set of importance weights and sensitivity analyses as discussed above revealed that Sexbierum stands out in nearly all evaluations as the site to be preferred. This forced order-of-priority matrix has proved to be a good aid in checking the results.

3. PROCEDURAL ASPECTS OF THE WINDFARM PROJECT

3.1 Introduction

On August 18, 1983 the Minister of Economic Affairs selected Sexbierum as the site for the windfarm project. The Minister's decision was also based on the advice from the Committee on Electricity Works dated 29 June 1983.

Part of this section will be spent on the procedural activities which eventually led to this advice. Attention will also be paid to the activities which had to take place on the site finally chosen, in order to obtain the necessary licences in time.

3.2 The National Evaluation Procedure

3.2.1 Introduction

It was considered highly important that the choice of the site for the windfarm was made very carefully. Especially because of the fact that this windfarm project might be the first step towards further utilization of wind energy in the Netherlands, it was essential that the procedures for this project met with the least possible hindrance from inaccuracies in the procedures to be followed. Therefore, the Provincial Planning Boards, and in some cases also the municipalities, were involved in the consultations at an early stage, via the Provincial Electricity Companies. To guarantee that the procedure was followed accurately, it was also considered important that an evaluation was made of the three sites finally proposed (Sexbierum, Noordoostpolder and Wieringermeer 2) on the basis of the site exploration.

For the national evaluation of these three sites the expertise of the Committee of Electricity Works (CEW) was called in. CEW's composition is such that a wide range of areas of special interest is covered. CEW advises the Minister of Economic Affairs with respect to the physical planning aspects of electricity works in general. Because of its experiences with electricity supply, and in view of the multidisciplinary expertise of its members, it was logical to let this committee play an important role in the national evaluation of the sites.

3.2.2. The CEW Procedure

Prior to the actual CEW procedure, a preliminary consultation was conducted, in which the following parties were involved: the municipalities concerned, the provinces concerned, the Agricultural Board (the national and regional boards concerned), the National Gas Board, the Department of Public Works, the Dutch Railways, the Ministry of Defence, the Civil Aviation Authority, the Historic Buildings Council, the National Planning Department, the Association for the Preservation of Natural Reserves, the Dutch Association for the Protection of Birds, the Inspectorate for Physical Planning in the provinces concerned, the Telephone Districts in the provinces concerned, District Water Boards, District Polder Boards, Environmental Protection Organizations, and the Domain Boards.

All opinions expressed by these bodies involved in the preliminary consultations were brought to the attention of CEW. In this way CEW could give due weight to these reactions in giving its advice (see Section 3.2.3). The national (integral) evaluation took place between January and May 1983. In the public hearings, which formed a part of the CEW procedure, persons and organizations concerned, as well as those who were interested, could express their opinions and objections before CEW. The hearings were preceded by information meetings organized by Sep. These information meetings and hearings were all held in April and May 1983.

During the hearings many objections were expressed especially by farmers and their organizations. These objections mainly concerned:
- the expected hindrance for agricultural exploitation of the lots where the windfarm would be built. The expected hindrance was thought to be greater in the case of more intensive exploitation of the lots concerned, as well as in the case of more mechanized cultivation;
- the damage caused by interested visitors;
- noise pollution and cast shadows from moving rotor blades;
- the fear of being used as Guinea pigs; and
- the fact that some aspects of the windfarm project were still unclear.

The latter two objections can never be fully remedied, because we are dealing with a unique project. The project consists of a windfarm with wind turbines of a capacity not found in a park configuration anywhere else in the world.

The agricultural organizations put forward suggestions to build the windfarm on sites designated as non-agricultural areas or on agricultural areas which have not yet been granted for agricultural purposes, for instance: the new Flevopolders or fallow industrial areas. The objection to these locations lies in the fact that the physical planning experiment could not be performed, and it is precisely this experiment which formed one of the main objectives in the site selection procedure. Only a location which is typical in both physical planning aspects and procedure will give an accurate reflection of the social acceptance of wind energy in the Netherlands.

Referring to the possibility of building windfarms in fallow industrial areas, the following should be taken into account. In the site selection procedure for the windfarm, various industrial areas were examined during the first selection. All these industrial areas were rejected because of the surface and roughness requirements related to the research character of the windfarm project, and for economic reasons. In relation to cost aspects a multi-functional use of the areas was found to be essential. The rigid pattern of the windfarm and the corresponding roughness requirement were in conflict with the requirement of multi-functionality.

Furthermore, any possible objections from agricultural organizations had already been taken into account in the site check, by giving the agricultural aspects high priority in the weighting procedure. Finally, agriculture was given a very high weighting-value in the sensitivity analysis (the "Beets version").

The techno-economic requirements of the project were defined in such a way that objections put forward during the petition procedure could either be solved or reduced to an acceptable minimum.

3.2.3 The CEW Advice

On the basis of reactions from persons and organizations concerned and those interested, as well as the Government's point of view in "Wind Energy and Storage" and the results of the preliminary consultations, CEW submitted its advice to the Minister of Economic Affairs (May 1983). In this advice CEW joined Sep in its preference for the site at Sexbierum. In CEW's considerations relating to the choice of the site, the following aspects had received special attention:
- the favorable position of the site at Sexbierum in relation to research aspects, special wind velocity and ornithological aspects;
- the relatively favorable agricultural ratio (stock farming/arable farming);
- the possibility of integrating the windfarm into a re-allotment project; and
- the regional economic policy.

In its advice CEW only took into account the immediately required area of the windfarm (60 ha). The possible extension up to 250 ha near the Sexbierum site was not taken into account, although Sep, in its "Proposition for the Site of the Windfarm Project", indicated that all eight sites selected were tested on this criterion and were found suitable for an extension of this kind. CEW held the opinion that it was necessary to start an entirely new procedure in case of a possible extension up to 250 ha. CEW attached great value to the fact that, with regard to possible future extensions of the park and the height of the wind turbines, and with regard to possible effects, there was still insufficient certainty.

Furthermore, CEW expressed its opinion that a compensation

settlement should be realized. In this settlement adverse effects on agricultural exploitation, which had not been taken into account, could be adequately compensated for. In addition, CEW stated in its advice that an extensive and thorough research programme should be started in relation to the planning and environmental aspects of wind energy. CEW restricted its advice to the site at Sexbierum. In regard to the other sites where the procedure had been started (Noordoostpolder and Wieringermeer 2) CEW did not give any judgement. In case the windfarm would not be built at the Sexbierum site, CEW would be prepared to consider the other locations at short notice.

On August 18, 1983, the Minister of Economic Affairs agreed to the construction of the windfarm at the Sexbierum site. The Minister was also asked for his approval of a power line connecting the windfarm and the substation at Herbayum. In this request CEW was again asked for its advice concerning the physical planning aspects of the power line. Overhead lines as well as underground alternatives were brought into procedure. After the public hearings, CEW advised the Minister for Economic Affairs to grant permission to build an underground connection. On July 26, 1984, the Minister did so.

3.3 Regional and Local Procedures

The approval of the Sexbierum site by the Minister of Economic Affairs was the start of further procedural talks with the Province of Friesland and the Municipality of Barradeel, in which the windfarm would be located. On a provincial level, a revision of the Friesland Regional Plan was necessary. Prior to the definitive decision by the Minister of Economic Affairs, the province of Friesland had already started a procedure for a partial revision of the Friesland Regional Plan. The Provincial Planning Board approved this revision and on June 29, 1983, the Friesland Provincial States also gave their consent.

On a local level, the Municipality of Barradeel stated at an early stage that it was prepared to give its full co-operation in obtaining the necessary licences quickly. On June 28, Barradeel took a "preparatory decision". This could only be realized after Sep had guaranteed that overhead lines as well as underground alternatives between the windfarm and the high-voltage substation at Herbayum could be brought into procedure. Furthermore, the Municipality wanted to have more certainty regarding the original extension possibilities within the 1 x 2.5 km area wanted by Sep.

Regarding this aspect Sep stated that any possible extension of the windfarm would only be effected after the windfarm project had demonstrated that wind energy was an acceptable option in the Netherlands. After the demonstration project would have come to an end, a possible extension would offer greater freedom of configuration of wind turbines, which would make it easier to fit into the existing patterns of land use. Soon after this Sep reaction the Municipality of Barradeel altered its Local Zoning Plan in such a way as to make the windfarm project possible.

Apart from a revision of the Local Zoning Plan, requests were made for a Building Licence and a Nuisance Act Licence for the project. The request for the latter was made by Sep on November 25, 1983. The objections put forward referred to the impact on the environment, traffic hindrance, noise pollution, danger for the surroundings, horizon pollution, interference on radio and TV signals and depreciation of houses.

For noise pollution a "provisional" arrangement was made, which had to be worked out in the future after a period of gaining some operating experience. The Nuisance Act Licence was granted by the Municipality of Barradeel on August 28, 1984. The request for a Building Licence was made on December 5, 1983, and the Licence was granted on March 26, 1984, by the Municipality of Barradeel.

3.4 Right of Way Agreements

In its advice concerning the site of the windfarm CEW felt it desirable that a compensation settlement should be made. In this settlement adverse effects on agricultural exploitation could be adequately compensated for.

In June 1983 consultations were started with the representative committee of the farmers concerned, in which the extent of the compensation settlements was discussed. In March 1984 both parties came to an agreement and the first compensation settlements were made with individual owners and users of the lots in and around the windfarm site. No special problems arose in this process.

4. EVALUATION

Among the Dutch public, especially among environmental protection groups, there is a highly positive attitude towards the use of sustainable and clean energy sources such as solar energy, hydro-power and wind energy. Wind is especially seen as a promising source of energy. This positive attitude has undoubtedly contributed to the fact that the selection process, which eventually resulted in Sexbierum being chosen as the site where the future experimental windfarm would be established, was not felt to be controversial.

The site-exploration method used met with little opposition. The question whether a different method might also have "worked" can not be answered. It is felt that, given a positive attitude towards wind energy of all those concerned, any method would have been adequate. Nevertheless, as has been stated in Section 2.4, the multi-criteria analysis was adopted because of the nature of the information it could provide concerning the various sites. In the multi-criteria analysis it is possible to compare qualitative information on sites and to apply arithmetic techniques for calculating overall preferences. This is what makes this method eminently suitable for dealing with questions such as the choice of a site as dependent upon a variety of criteria.

On a provincial and local level, many aspects played a role in the decision about where the windfarm would be realized, such as publicity, employment and tourism (the windfarm could well become a tourist attraction in the near future). Thus consultations with various authorities proved to be necessary and useful. All this resulted in the fact that the licences requested (concession from the Minister of Economic Affairs, a revised Regional Plan, an altered Local Zoning Plan, a Nuisance Act Licence and a Building Licence) were obtained without delay.

The conclusion of the Right of Way Agreement, brought about in mutual consultations with those directly concerned, proved to be an essential requirement for obtaining the remaining licences.

REFERENCES

Begeleidingscommissie Voorstudie Plan Lievense. (1981). *Windenergie en waterkracht (Wind energy and hydro-power)*, 's-Gravenhage, mei.

Keeney, R.L. and Raiffa, H. (1976). *Decisions with multiple objectives: preferences and value trade-offs*. New York: Wiley.

N.V. Sep (1983). *Locatievoorstel proefwindcentrale (Siting proposal experimental windfarm)*, Arnhem, januari.

Roy, B. (1968). Classement et choix en présence de critères multiples (La méthode ELECTRE 1). *Revue Informatique et Recherche Opérationelle, 8,* 57-75.

Voogd, J.H. (1980). *Multi-criteria-methoden voor ruimtelijke evaluatie-onderzoek (Multicriteria methods for evaluation in physical planning).* Planologisch Studiecentrum TNO, Delft.

Tweede Kamer der Staten Generaal. (1982). *Windenergie en opslag (Wind energy and storage),* zitting 1981-1982, 17.500, nrs. 1-2, 's-Gravenhage, juli.

Zeleny, M. (1982). *Multiple criteria decision making.* New York: McGraw-Hill.

CURRENT RADIOACTIVE WASTE MANAGEMENT POLICY IN THE NETHERLANDS

HENK C.G.M. BROUWER

Division of Environmental Impact Assessment
Ministry of Housing, Physical Planning and the Environment, Leidschendam

1. INTRODUCTION

Like many other industrialized countries the Netherlands has had (limited) streams of radioactive wastes for a few decades. The management of these wastes has become a major political issue in the last ten years mainly under the pressure of public concern. This chapter deals with recent developments in the Netherlands in the field of radioactive waste management.
Two questions are crucial in this context:
- 'Where' shall we put it?
- 'How' shall we do it?
The chapter offers information on the formulation of answers to these two questions in a more or less chronological order.

In this chapter the term "Government" is frequently used. For a better understanding of this term the reader must be aware that in the Netherlands the basis for taking decisions on "nuclear issues" lies in the Nuclear Power Act of 1963. This Act combines the promotion of nuclear power and the protection from radiation. It designates the Minister of Economic Affairs as the first responsible minister for decisions and furthermore requires agreement from the Minister of Social Affairs, the Minister of Housing, Physical Planning and the Environment and the Minister of Health. In practical politics (Parliament), the Minister of Economic Affairs represents the Government for energy production matters, whereas the Minister of the Environment is usually the spokesman for all other aspects (safety, waste management). The largest civil service organisation for nuclear issues is part of the Ministry of Social Affairs (reactor safety in view of workers' protection).

Ch. Vlek and G. Cvetkovich (eds.), Social Decision Methodology for Technological Projects, 217–234.

2. PREVIOUS POLICY ON RADIOACTIVE WASTE

2.1 Ocean dumping of low and middle level waste

For many years the low and middle radioactive wastes from, e.g., laboratories and hospitals and the existing nuclear power stations were collected and transported to a site at the Netherlands Nuclear (now Energy) Research Foundation in the dunes of Petten (municipality of Zijpe). On this site the wastes were packaged and stored for a short time. Then they were shipped in the Velsen-harbour for dumping into the Atlantic Ocean. The dumping operations took place from 1965 onwards.

In the early years nobody worried about this method of waste disposal. However, from the middle of the 1970s, when the Dutch nuclear energy debate really got going, environmental and other groups began to resist the ocean dumpings. Finally, the bi-annual shipping in Velsen was accompanied by demonstrations and roadblocks of 'human chains'. As a result, the Dutch Government (i.e. the Minister of the Environment) decided to stop the dumping operations in 1982 and to apply the same philosophy to these radioactive wastes as to other wastes: isolation, control, and supervision.

2.2 Intermediate land-based storage

The decision to stop the ocean dumpings was made under public and political pressure. However, this decision did not mean that a storage site on land was available. Such a site was necessary as the facility in Petten was only designed for short term (two years) storage. Therefore, an attempt was made to get an alternative site very quickly. The approach taken was based on the firm belief that a storage facility would not lead to any technical, safety or environmental problems (these issues were to be covered by a licence from the Government). Hence, it seemed only necessary to find and buy a site without restrictions in the land use plan for the area. In that case the local authorities would be legally obliged to grant a building licence. This also meant that it was not necessary to discuss the matter in advance with local authorities.

In November, 1982, a "suitable" site was identified in the industrial area of the municipality of Velsen. At the same time a new Minister of the Environment came into office. An information package for the public was prepared by the Ministry of the Environment. On the evening of the day before 'the news' was to be officially announced the mayor of Velsen was informed. In the following hours the information-package was delivered at every house in the municipality.

This strategy turned out to be not very successful. The municipality and people of Velsen refused to accept the storage facility under all circumstances. Fortunately, Velsen and Zijpe (harbouring the short term storage facility) are both in the province of North-Holland. The Minister of the Environment then started consultations with the provincial authorities and the local authorities of Zijpe. In 1983 an agreement was reached on a limited expansion of the short term storage facility in order to get time for the Government to find an alternative solution. The Government committed itself to use the expanded facility no longer than five years (under special circumstances ten years) and to make a decision on a new site before 1 January, 1986. To compensate the local community, the Government promised a cut in military operations in the area.

2.3 Previous policy on high radioactive waste

High radioactive waste originates from nuclear power stations. It mainly consists of spent fuel elements. The Netherlands has two nuclear power plants at Borsele (1973, 500 MegaWatt electricity) and Dodewaard (1968, 55 MWe). Since the mid-1970s there has been a public debate on the final disposal of high radioactive waste. A salt dome is considered to be the only feasible method in the Netherlands. However, in the areas where such salt domes are available there exists a very strong public and political opposition. This opposition has resulted in a parliamentary moratorium on salt dome drillings. Therefore, it is impossible to obtain any further information on the risks of this option.

In 1978 contracts were made with Cogéma (France) and British Nuclear Fuels to reprocess spent fuel elements from Borsele and Dodewaard, respectively. These contracts leave open the possibility for the reprocessing plants to return to the Netherlands the waste from the reprocessing or the unprocessed fuel elements themselves. These returns may begin in 1992.

3. TOWARDS INTEGRATED RADIOACTIVE WASTE MANAGEMENT

From the above it is clear that by 1983 the Dutch Government was obliged to take further action on radioactive waste management. For low and middle radioactive waste there only was the temporary (maximum ten years) solution at Petten, whereas for high radioactive waste there was no solution at all, in case the reprocessing plants would return waste. For the latter problem a ministerial committee was formed. This committee investigated the option of storage of high radioactive waste above ground level. It came to the conclusion that both water-cooled and air-cooled storage would be technically possible. Furthermore it was decided that there should be one single organisation for the implementation of radioactive waste management policy: COVRA (Central Organisation for Radioactive Waste Management).

This radioactive waste management policy was laid down in a *White Paper* to Parliament in April, 1984. The White Paper described the waste problem in quantitative terms based on the existing production of radioactive waste. It stated that no serious environmental effects were to be expected from a storage facility. The most important message from the Government was the announcement that there was going to be one facility for the long-term interim-storage (50-100 years) of all radioactive waste. For the preparation of the site selection for this facility the Minister of the Environment would appoint a special committee (the LOFRA-committee). The final selection would be made by the Government itself (before 1 January, 1986, as promised to the municipality of Zijpe).

In October, 1984, Parliament discussed the White Paper with the Minister of the Environment. The majority agreed with the proposals of the Government.

3.1 The LOFRA site selection - initial stage

The LOFRA-committee had to answer the question of 'Where shall we put it?". As the Velsen experience of 1982 had shown, it would be crucial to find a cooperative local authority. Therefore, the main function of the LOFRA-committee was to investigate the willingness of local authorities to negotiate with the Minister of the Environment and COVRA. The Minister of the Environment appointed the committee in December, 1984. Soon it became known as the Geertsema-committee after the name of its chairman. The five committee members were chosen on their personal quality and reputation. Mr. Geertsema was a former Minister for the Interior and had been the Queen's Commissioner in the province of Gelderland. Three other members of the committee were (former) members of the Privy Council (Raad van State), the most important advisory body to the Government on legislative issues. In fact the whole committee consisted of ex-politicians from different parties.

The official task of the committee was to submit a report to the Minister of the Environment in which about three possible sites would be presented. This report was expected by 1 October, 1985. The committee carried out an initial desk study in December 1984 and January 1985. The following selection criteria were used:
- the site should be an industrial area according to land use plan,
- it should cover at least 75 acres of land, and
- there should be a possibility for discharge of cooling and processing water.

This exercise led to twenty initial candidate sites distributed throughout the country. In February 1985, the committee sent a letter to all provincial (not local) authorities with the request to check the information on the pre-selected sites in their area. Within a few hours the media had found out about all twenty sites. The (negative) reactions from the public and the authorities (though not all of them) were hot news in all media.

The committee continued its desk study taking into account the information that the provincial authorities had provided upon the committee's request. In April, 1985, an interim-report was submitted and published. The number of sites was reduced to twelve. Then the committee started consultations with provincial and local authorities in order to find out about their willingness.

3.2 Environmental Impact Assessment - initial stage

The LOFRA-committee was only established to deal with the 'where'-question. The committee had no mandate, nor expertise on 'how' the facility was to be constructed and operated, or about its risks. In order to fill this 'how'-gap it was decided in January, 1985, that an Environmental Impact Assessment (EIA) procedure would be set in motion. In the Netherlands such a procedure entails the involvement of an independent expert panel (EIA-commission). An Environmental Impact Statement (EIS) must be submitted by the proponent of an activity (in this case COVRA). Strictly speaking, the EIA procedure could have been initiated by COVRA *after* a site-selection had been made, i.e., while applying for the licence required under the Nuclear Power Act (NPA). However, in this case COVRA made a request for guidelines for its EIS in February, 1985. The Minister of the Environment then decided to carry out a two-phase assessment of the potential effects on the environment:

    (1) a general, site-independent EIA, to be submitted by October, 1985;

    (2) a more specific, site-related EIA, to be submitted together with the application for the NPA-licence, by the end of 1986.

In March, 1985, COVRA repeated its request for guidelines for the (site-independent) EIS. It also produced an initial paper which contained information on the envisaged facility for different scenarios of nuclear electric power generation. The general public could obtain a free copy of this initial paper as well as an explanatory note on the procedure (the combination of the 'where' and 'how' lines of investigation). The relevant provincial and local authorities were directly provided with this information package. The presented procedure made it clear that the Minister of the Environment would not make a final site selection (based on the LOFRA advice) until he would have received the EIA-commission's review of COVRA's Environmental Impact Statement.

During March and April, 1985, the public, and the local and provincial authorities had, and took, the opportunity to express their views on the issues that would need attention in the EIS. In May, 1985, the independent EIA-commission submitted its advice to the Minister of the Environment in the form of draft guidelines for the EIS. These draft guidelines suggested, inter alia, that COVRA:

    - provide details on the design of the facility;
    - separate the storage of low/middle from high radioactive waste;
    - also consider decentralized storage (i.e., at the site of existing nuclear power plants); and
    - consider attitudes and possible reactions of immediately concerned groups of citizens.

It may be noted that the chairman of the EIA-commission, dr. J. Spaander, was the founder and former director-general of the National Institute for Health and Environmental Protection. Other members were experts in the field of environment, nuclear safety, health and psychology. One of the experts was from the Federal Republic of Germany (Federal Health Institute, Berlin).

3.3 Interferences and complications

While the first steps were taken towards the realisation of a single storage facility for all types of radioactive waste, in January, 1985, the Dutch Government announced its intention to increase the country's nuclear power generating capacity with 2.000 - 4.000 MWe in the next decade. Of course, in comparison to the White Paper of April, 1984, this increase meant a significant change in the amounts of radioactive waste that could be expected in the future. The consequences became clear in the initial paper that COVRA submitted in the EIA procedure in March, 1985. They were incorporated in the LOFRA site-selection criterion of 75 acres.

In the beginning of June, 1985, a really serious disturbance came from an unexpected side: other Ministries of the same Government. It turned out that, probably after secret negotiations, a few Ministers and Secretaries of State (Interior, Treasury, Economic Affairs) had put a kind of obligation, in writing, on the Moerdijk Industrial Estate Authority (South of Rotterdam) to accept the radioactive waste in exchange of a good price for 75 acres of its property. This letter contained a scheme of (this part of) the Government to relieve the financial trouble which this Authority had incurred. The Moerdijk area was one of the remaining twelve candidate sites at the time. Neither the LOFRA-committee nor the Minister of the Environment were informed about this action. Although the Minister of the Environment, in consultation with the Prime Minister, immediately assured Parliament that this letter to the Moerdijk Authority did by no means influence the site selection procedure, two members of the LOFRA-committee decided to quit. The reactions of the public and the media were very negative. There was a strong impression that the selection had already been made in a deal behind closed doors and that the LOFRA work was only window-dressing.

A third complication came by the end of June, 1985. At that time, the Dutch Parliament discussed the Government's proposal to increase

Table 1:Timetable of events relevant for the radioactive waste storage debate

1968	Dodewaard nuclear power plant (55 MWe) in operation
1973	Borssele nuclear power plant (500 MWe) in operation
1968	Radioactive wastes stored at Petten for bi-annual ocean dumping
1974	Government proposes three new nuclear power plants; start of a long debate on nuclear power in the Netherlands
1978	Reprocessing contracts with BNFL and Cogéma, including return clause
1978-81	Growing protests against ocean dumping
1981-83	Formally conducted nationwide discussion on future (nuclear) energy policy
1982	Government decides to stop ocean dumping
1982	Government's attempt to establish storage facility in Velsen fails
1983	Temporary arrangement (5-10 years) for storage at Petten
April 1984	White Paper to Parliament: one land-based interim-storage facility for 50-100 years for all levels of radioactive waste
Oct 1984	Parliament approves White Paper approach
Dec 1984	LOFRA (site-selection) Committee established for storage facility
Jan 1985	Government proposes 2-4 new nuclear power plants
Jan-March 1985	Start of site-independent EIA by COVRA (involving public and independent expert Commission)
Feb-April 1985	Desk study by LOFRA (from 20 to 12 sites)
May 1985	Advice of independent EIA Commission on guidelines for EIS
May-Oct 1985	Consultations of LOFRA with provincial and municipal authorities (from 12 to 2 sites)
June 1985	Publication of a letter of some Government ministers promising one municipality to 'get' the storage facility
June 1985	Parliament approves in principle of the scheme for new nuclear power plants on the condition that a solution be found for the radioactive waste.
Sept 1985	LOFRA committee completes final report recommending two sites
Nov 1985	Site-independent EIS submitted by COVRA
Jan 1986	Site-independent EIS severely criticized by independent EIA-commission of experts
April 1986	The nuclear reactor accident at Chernobyl, USSR
June 1986	COVRA decides on selection of Borsele as best storage location
July 1986	Start or work on site-dependent EIA by COVRA
Nov 1987	Publication of site-dependent EIS
Jan 1988	EIA-commission judges site-dependent EIS to be a "sufficient document" for further decision making on the facility

nuclear power facilities with 2000 - 4000 MWe. The Parliamentary majority agreed to this proposal, but only under the condition that an acceptable solution for the radioactive waste problem (i.e., the storage facility) would be available. At the time this was explained as a "grantable" licence. At any rate, the now explicit link between the construction of new nuclear power stations and the waste storage facility would put pressure on the procedures for the storage facility in the future, as the energy-production sector had its own deadlines for procedures and dates of starting construction activities.

Table 1 gives a chronological summary of events relevant to the debate on the storage of radioactive waste.

3.4 The LOFRA - consultation and report

As stated above, in April, 1985, the LOFRA-committee began consultations with provincial and local authorities of the twelve remaining sites with only one question: "are they willing (enough)?". These consultations lasted until August, 1985. Local and national media gave extensive coverage to the committee's tour of the country. There were some small scale demonstrations and a few controversial statements by the chairman on the safety issue ("the storage facility is less risky than a semolina factory"). Every meeting was followed by a press conference where both the committee and the authorities involved answered questions.

In the twelve municipal councils serious debates were held on the position each municipality should take. In these discussions the 'how' question was often addressed as well. Unfortunately, these 'how' discussions had to be based on limited information, i.e., the initial paper by COVRA and information professional environmental groups had sent to the members of the municipal councils concerned.

During the month of September, 1985, the LOFRA committee completed its report which was presented to the Minister of the Environment on 8 October, 1985. The committee recommended two sites for the storage facility:
- Borsele, the municipality already having a 500 MWe nuclear power plant within its boundaries; and
- Moerdijk (in the municipality of Klundert, South of Rotterdam) with its financial troubles.
Both sites are part of large industrial estates.

Near the end of the consultation period a third alternative site appeared through an offer by the Provincial Electricity Company of Zeeland (PZEM), the owner of the Borsele nuclear power plant. This offer concerned an area quite near the power plant. At the same time both Borsele and Moerdijk were also in the race for one or more of the new nuclear power plants planned to be constructed in the near future.

The committee did include the third option in its report but had no time to really investigate its suitability. In the opinion of the committee, the first-mentioned Borsele site was slightly preferable to the Moerdijk option. According to the committee, Borsele was preferable because the local authorities had shown a greater political willingness. The other ten municipalities, the committee reported, had refused to accept the storage facility for psychological and political reasons which were based on anti-nuclear and environmental grounds.

With the submission of its report the LOFRA-commission had accomplished its task. The Minister of the Environment immediately submitted the report to Parliament stating that a decision on the site would be made in February, 1986. In the meantime, the Minister stated, he would consult with the two remaining local authorities in November, 1985, and decide after COVRA's site-independent EIS had been reviewed, particularly by the independent EIA-commission.

3.5 The EIA guidelines

After public comments and the advice (draft guidelines) from the EIA-commission in May, 1985, the Minister of the Environment issued guidelines for the EIS that COVRA was to prepare. The Minister of the Environment did not decide on the contents of these guidelines only by himself. There was also input from the Ministries of Economic Affairs and Social Affairs.

The guidelines were published in July, 1985. They required a fair amount of detailed information and some attention was to be paid to psychological aspects. However, they did not emphasize possibilities for the separate storage of low/middle and high radioactive waste nor for a decentralized storage. In an annex to the guidelines the Ministers explained why they had not followed some of the suggestions from the public or the EIA-commission.

3.6 The COVRA plan - preliminary report

In its site-independent EIS COVRA presented its plans for dealing with radioactive waste in general. Two scenarios where used for the estimation of the quantities and qualities of future radioactive wastes in the next hunderd years:

- *scenario 1*: the existing situation (two small nuclear power plants) extrapolated into the future (including return of high radioactive wastes from BNFL and Cogéma).
- *scenario 2*: scenario 1 plus the waste from new nuclear power stations with a capacity of 2000 MWe.

The most important part of COVRA's plans pertains to high radioactive waste that needs cooling. The EIS describes the way of transportation from the various sources as well as the initial handling of this waste upon arrival at the storage facility (in a 'hot cell' the waste is loaded from transport containers into storage containers or cylinders). In scenario 1 COVRA suggests storage in iron-cast containers with air cooling. The containers should provide protection from radiation and against external influences. In scenario 2 COVRA suggests storage in cylinders in a so-called vault (a concrete bunker) also with air cooling. This bunker should provide protection from radiation and against external influences. In both scenarios there will be a permanent monitoring program on possible leaks in containers or cylinders.

In an alternative plan (not preferred by COVRA) the possibility of a vault for fission waste cylinders in combination with a water basin for the storage of radiated fission elements is described for scenarios 1 and 2. Moreover the alternative of container storage in a normal building (not a vault) is described for scenario 2. In a second alternative plan all high radioactive wastes are stored in containers in a normal building.

Under normal operations the radiation from the alternatives for the storage facility will only be a fraction of the natural radiation level, the EIS states. COVRA also investigated a number of possible accidents and quantified the radiological consequences of the most serious accidents. The worst accident would be an air plane crashing on a loading machine filled with radiated fission elements. In its EIS COVRA concludes that in such a case the radiation dose outside the storage site would be in the same order as the yearly individual radiation level from natural sources.

3.7 The EIA review phase

In the beginning of November, 1985, i.e., one month after the publication of the LOFRA-report, COVRA submitted its site-independent EIS. Within the Ministry of the Environment there was too much time pressure to consider the quality of this EIS seriously. Moreover the eventual rejection of the COVRA report would have had an immediate impact on the site selection procedure (the LOFRA follow-up). It may be recalled that Parliament had linked the storage facility to the future construction of new nuclear power plants. Thus, the COVRA EIS was published within ten days and submitted to Parliament. The public and governmental advisors were given the opportunity to make comments during one month, while the EIA-commission got an extra month to complete its review. Very soon the EIS was severely criticized, not only by the public (mostly the environmental groups) but also by advisors (e.g., the Chief Inspector for Environmental Protection), and - most significantly - by the independent EIA-commission.

At the end of January, 1986, the EIA-commission submitted its review to the Minister of the Environment. The commission's unanimous advice stated that the EIS treated the envisaged storage of low and middle radioactive waste sufficiently. However, the way in which high radioactive waste was to be stored, the risks involved and the protection measures to cope with these risks were described insufficiently and inaccurately.

Therefore, the EIA-commission advised the Minister for the Environment to have COVRA improve its case in a second, site-related EIS. This EIS should cover all three remaining sites (Moerdijk and Borsele 1 and 2) as the reviewed EIS was an inadequate basis for a final site selection. The commission added to its review that its severe criticism of the EIS did not mean that its members thought that an environmentally safe storage facility would be infeasible. As usual, the commission's review received quite some attention in the media.

In addition to the review the EIA-commission also presented its advice on the guidelines for the site-related EIS that COVRA must submit in conjunction with the application for a Nuclear Power Act Licence. In this advice the EIA commission indicated several issues that must be dealt with separately for each individual site (e.g., risks ensuing from nearby chemical plants).

3.8 The LOFRA final site selection

From the consultations of the Minister of the Environment with the local authorities of Borsele and Moerdijk (Klundert) in November, 1985, the conclusion was drawn that there were no major differences among the three possible sites. Particularly, after the authorities of Moerdijk had dropped some of their original reservations both municipalities could be considered equally eager! The rather negative results of the site-independent EIA did not change their attitude. In this situation the Minister of the Environment decided to give up the original plan to make the final site selection himself.

On the eve of the parliamentary debate on the site selection of new nuclear power plants a note was sent to Parliament on 21 April, 1986. It announced that the choice was left to COVRA until 1 July, 1986. This deadline was put on COVRA, because otherwise the storage facility licence procedure would delay the licensing procedures for the new nuclear power plants. In view of this same objective, COVRA was to submit its site-related EIS (and licence application) by January, 1987. Moreover the note to Parliament announced that COVRA in its EIS should cover the site-specific aspects for all three sites (regardless of its choice) and provide all missing information as pointed out by the EIA-commission. The guidelines for the EIS would soon be published.

As indicated above Parliament was to discuss the Government's proposal on the site selection of new nuclear power plants. In this proposal the Government repeated its preference for Borsele. The Parliamentary debate began on 21 April, 1986. By means of a formal procedural manoeuvre the opposition parties managed to postpone discussions for a week. Within that week, on 26 April, 1986, the Chernobyl disaster started to unfold. In the beginning of May the Dutch Government decided to postpone the decision process on any new nuclear power plants until a thorough evaluation of the Chernobyl accident had been accomplished. Parliament welcomed this decision (three weeks before a general election) and also accepted that COVRA was to select its storage site.

The decision to postpone the construction of new nuclear power plants could have allowed COVRA to take more time for its site selection for the storage facility, because the majority in Parliament did not need a grantable licence for the waste storage anymore. However, COVRA presented its choice on 18 June, 1986: the site of the Provincial Electricity Company of Zeeland at Borsele.

The Chernobyl disaster has caused some delay in the issuance of guidelines to COVRA for the site-related EIS. These guidelines were published in late June, 1986. A second consequence of Chernobyl is that the time pressure on COVRA to produce its site-related EIS and licence application had diminished. These documents were actually submitted in the Fall of 1987. See Table 1 (section 3.3) for an overview of the chronological order of events around the waste storage problem.

4. SOME OBSERVATIONS ON DECISION METHODOLOGY

In the above an attempt has been made to present factual information in a more or less chronological order. It is a description of a decision making process with two main questions about the storage of radioactive waste: 'where?' and 'how?'.

The 'where?' question was considered to be the most difficult one to answer, particularly after the experience at Velsen. Therefore a special independent committee was formed (LOFRA), consisting of highly respected ex-politicians. This committee operated in full daylight and started an open selection process. It consulted provincial and local authorities (not the public). It used a limited amount of technical data. It did not get into bargaining or behind-closed-doors operations (however, some parts of the Government did or made the impression to do so). It successfully, and in the end rather easily, accomplished its task.

Thus, the 'where?' problem seems to have been solved without any force on the part of the Government. Whether or not this is due to the activities of the LOFRA committee, no one can tell. Some may think that the two municipalities (Borsele and Moerdijk) which were finally contesting the storage facility can not be considered as average municipalities. Others may emphasize the fact that two volunteers showed up against everybody's expectation.

The 'how?' question was considered to be a solved problem from the very beginning. In fact, the LOFRA-committee presented the 'how?' question as "technical details to be dealt with in a later procedure". This can be explained in connection with the Government's White Paper of April 1984. Notwithstanding this White Paper the Minister of the Environment initiated a separate procedure on 'how?'. This was the site-independent Environmental Impact Assessment (to be followed by a site-related EIA). This procedure put strong information requirements (to provide technical data) on COVRA. The procedure also allowed for some involvement of the public. Moreover an independent experts' commission played an important role.

The initial EIA procedure had by no means led to a solution of the "how?" problem. However, this did not become clear until after the submission of the LOFRA-report. Fortunately, there was a second opportunity for COVRA in the site-related EIA to demonstrate that a safe storage facility for all types of radioactive waste could be constructed and operated.

Finally, one may note that the LOFRA and EIA approaches could not be applied as blueprints. This was due to several related decisions, developments and incidents inside and outside the Government. This real life case story may, therefore, also demonstrate the relativity of any (set of) formal social decision methodology(ies).

EDITORIAL POSTSCRIPT

In January 1988, with the EIA-commission's positive evaluation and the ministerial acceptance of COVRA's voluminous site-dependent EIS for the Borsele location, the story was not ended yet. Following the administrative procedure the municipal council of Borsele had to consent with the requested licence for COVRA to construct the storage facility. During the month preceding the planned council meeting of March 1, 1988, several people of Borsele founded a 'village council' which quickly distributed a questionnaire and polled local public opinion about COVRA's plans. The response was an overwhelming "no" accompanied by a strong majority judgment that the local government had failed to inform its citizens adequately and to represent their interests vis-à-vis the provincial and national authorities.

There are several factors which prompted the Borsele people to undertake this concerted action. First, publication of COVRA's second, site-dependent EIS in November 1987 had elicited several hundred statements-of-objection among organizations and inhabitants of the region. Apart from the expected concern about ionizing radiation from the facility, many writers objected to the further 'nuclearization' of the Borsele area and feared a national public stigmatization of themselves and especially their agricultural and dairy products as being 'radioactive'.

Secondly, representatives of COVRA and of the Minister of the Environment had participated in several meetings organized to inform the local population and to offer them an opportunity for asking questions. There it gradually became clear, to many visitors' distress, that the planned facility would come to lie much closer to, and would be clearly visible from, the village than they had thought all along.

Third, the EIA-commission in January 1988 had indeed judged the site-dependent EIS to be an acceptable policy document, but it had not done so uncritically. Based on the EIS itself but also on the many statements-of-objection from the region, the EIA-commission concluded that a major shortcoming - mostly on the side of the Minister of the Environment and the local authorities - had been the lack of sufficient and fully open information towards the various organizations and public groups involved.

During a crowded, noisy meeting on March 1, 1988, the municipal council of Borsele ruled that an interim committee supported by external experts was to search for an alternative location inside the Borsele area within two months. Such an 'at least as good and possibly better' location was eventually proposed; it is situated at about 1500 meters instead of the original 750 meters from the village, and it is surrounded by other industrial sites rather than that it directly borders on farmland and the village beyond. A peculiar detail is that the new location did not figure on the LOFRA search committee's original list of candidate sites.

Now, at the end of 1988, COVRA is preparing its third (largely new) site-dependent EIS following additional guidelines issued by the Minister of the Environment in November 1988. Again, the EIA-commission has advised the Minister to fully inform the local population in time and preferably also poll their (informed) opinions. This time a new element has been proposed: the formation of a local liaison committee which is to follow the whole (remaining) EIA and licensing procedure in close understanding with COVRA and the relevant authorities. The idea is that a liaison committee composed of local representatives from industry, agriculture and the general public may be instrumental in establishing a reliable communications basis and the mutual trust required to construct, operate and maintain the storage facility in a safe and otherwise acceptable manner, during the intended long period of its existence.

From a social decision-methodological point of view it may be concluded that many things have gone wrong during the four years that it was attempted to solve the problem of the long-term interim surface storage of all radioactive waste. First, for a novel and far-reaching project like this, a technical, economic and environmental analysis of its feasibility and possible effects is apparently insufficient. Apart from the project's initiator and its supporting national authorities there are various other groups of people involved. They want to know about and to have a say in the manner in which the project is designed, is decided upon and will eventually be managed and controlled. Living in a democracy one might even say that such groups have a right to know about the positive and negative possible consequences of a project of this nature, scope and expected duration.

Second, it would seem to have been better to start with the preparation of an environmental impact statement in an attempt to carefully answer the question of "how". Given this information a national search for an optimal location would have had better foundations and it could therefore have proceeded more efficiently and successfully.

Third, the original problem of where the Netherlands should store its existent and acceptably-produced radioactive waste was unwisely contaminated by the Government's principal decision of January 1985 to build several nuclear power stations in addition to the existing small plants at Dodewaard and Borsele. This decision made the waste storage project to be seen as an entrance door (some called it a crowbar) towards a significant expansion of nuclear energy use. In discussions about nuclear power and the radioactive waste problem quite a few people expressed anger about this 'polluted' way of policy making and enforcing critical decisions.

Finally we may note that, in the Netherlands, the law on

environmental impact assessment is fairly new, so that all parties concerned have had only a limited period of time and a limited number of cases for developing and accumulating experience. So far it is not clear, for example, what could, or should, be the role of formal multicriteria analyses in evaluating a proposed project and its alternatives (see Lubbers, this volume). Also, there is uncertainty about what would be a proper handling of public objections, protests, or just alternative views. And, of course, there are regular feelings that a significant lack of knowledge exists with respect to crucial points in the assessment of possible impacts.

Ch.Vlek[1]

REFERENCES

Lubbers, F. (this volume). Planning and procedural aspects of the windfarm project of the Dutch electricity generating board.

NOTES

1. The author has for some time been narrowly involved in the radioactive waste storage case. Thanks are due to P. Hermens of the Ministry's Division of Radiation Protection for his comments upon an earlier version of this postscript.

NUCLEAR WASTE: PUBLIC PERCEPTION AND SITING POLICY

JOOP VAN DER PLIGT

Department of Psychology
University of Amsterdam

1. INTRODUCTION

The siting of radioactive wastes poses a significant planning challenge to many countries. The public is generally extremely apprehensive about radioactive waste, and this has led to substantial delays in siting much needed waste facilities. As a consequence, the 'back end' of an industry that started some three decades ago still poses serious problems to its further development. Scientists continue to disagree about the best technical means for permanently isolating radioactive wastes from the biosphere. Moreover, most governments have not yet resolved many of the relevant political and institutional issues surrounding the waste management problem. The siting problems not only affect the nuclear power industry; the controversies of low-level radioactive wastes affect a wide variety of other nuclear-related industries, biomedical research facilities and hospitals (Welch, 1985).

This chapter presents a brief overview of the current situation of siting radioactive wastes. This is followed by an overview of various psychological approaches attempting to analyse public reactions to nuclear facilities. It will be argued that public reactions to nuclear waste facilities must be seen in the context of more general attitudes toward nuclear energy. The latter are not only based upon perceptions of the health and environmental risks but are built on values, and sets of attributes which need not be similar to the representations of the experts and policy-makers. The issue of siting nuclear waste facilities is also embedded in a wider moral and political domain. This is illustrated by the importance of equity issues in siting radioactive wastes. In the last section, the implications of the present line of argument for risk communication and public participation in decisions about siting radioactive wastes will be briefly discussed.

Ch. Vlek and G. Cvetkovich (eds.), Social Decision Methodology for Technological Projects, 235–252.
© *1989 by Kluwer Academic Publishers.*

2. SITING RADIOACTIVE WASTES

As early as the mid-1950s scientists recognized the necessity to develop safe methods for storing nuclear wastes. Nuclear energy generation leaves several dozen radioactive byproducts in "spent" fuel rods. This high-level waste can be handled in one of two ways, both involving waste siting decisions: (1) the rods can be transported directly to a permanent waste-management facility designed to keep them isolated from the biosphere for hundreds of thousands of years, and (2) they can be transported to a reprocessing plant where some elements (unconsumed U-235 and fissionable plutonium) are extracted and made into new fuel rods. The resulting wastes are then solidified and transported to permanent waste storage facilities. Low-level waste, such as that produced by medical facilities is directly deposited into waste-management facilities.

Since 1957 numerous scientific appraisals of the waste disposal problem have been presented to the relevant authorities with solutions ranging from salt mine burial of wastes, deep-well disposal, sea-dumping and various retrievable surface storage methods. In the mid-1970s the debate escalated. Various experts expressed their doubts about the safety of the waste management methods used by commercial operators. In the U.S.A., both the National Research Council and the Environmental Protection Agency recommended to postpone the nuclear program in order to study the safety of reprocessing facilities. Other scientists dismissed these proposals, saying that waste disposal was just "not a major problem".

The low level of agreement within the scientific community and the relevant authorities has contributed to the present stalemate. One of the consequences is that high-level radioactive waste in the form of spent fuel rods accumulates in cooling ponds at the individual reactor sites. The lack of progress in siting low-level radioactive wastes has similar consequences, not only for the nuclear industry but also for a variety of other industries, medical research facilities and hospitals.

Lack of public acceptance played an important role in the considerable delays in siting programs for both high- and low-level radioactive wastes. Siting efforts frequently encountered determined and vehement local opposition, not only in the U.S.A., but also in various Western European countries. The U.S. Environmental Protection Agency acknowledged the intensity of local opposition by observing that "siting efforts seem to unite grandmothers and U.S. congressmen, factory workers and university scientists, those who never graduated from high schools and those with doctorates in ecology and physical sciences" (U.S. EPA, 1979, iii).

3. PUBLIC OPINION

Public opposition can be traced back to the mid-1970s. Until then the debate on nuclear energy could be characterized as a technical debate about technical issues. Since then, public support for nuclear energy had gradually eroded. The press has shown substantial concern over nuclear safety and the possible consequences of a number of accidents (e.g., Harrisburg, Sellafield and Chernobyl) have been widely reported. This increase in media attention was accompanied by increasing public concern over potentially catastrophic accidents and radioactive wastes. Public concern is also reflected by various national referenda on the issue of nuclear energy (all decided by very narrow margins, e.g., in Austria, Switzerland, Sweden), and the results of a series of state-initiative votes in the U.S.A.

The many surveys conducted over the last decade provide a further illustration of the erosion of public support. Prior to the mid-1970s survey data showed consistently high levels of support for nuclear energy. The place of nuclear energy as a source of electrical power seemed assured. This majority eroded in the 1970s. Although this slippage was apparent prior to the Three Mile Island accident in 1979, it was accelerated by that event. There has been some rebound toward pre-Three Mile Island levels of support and opposition but this return has not been complete.

Recent figures show that the percentage of the U.S. public that supports the continued building of nuclear power plants in the United States is, on average, 5% to 10% more than the percentage of the public that opposes such construction. Furthermore, a majority of the public believes that more such accidents are likely to happen (Rankin, Melber, Overcast and Nealy, 1981). Finally, a substantial majority of the public (about 80%) now says that it is concerned about waste management issues (Kasperson, Berk, Pijawka, Sharaf and Wood, 1980). Yankelovich and Kaagan (1981) reported findings indicating that 80% of the U.S. public were of the opinion that the country had not progressed far enough on both disposal and transportation problems, and as a consequence, were "worried".

The above trends are also apparent in Europe and the U.K. Public opinion in the Netherlands has shown an "anti-nuclear" majority since the late 1970s. Opinion poll data for the United Kingdom show a slow but steady increase in public opposition to nuclear energy since the mid-1970s. Whereas in 1980 there was hardly any difference between the number of opponents and supporters of nuclear energy, a National Opinion Poll survey conducted in October 1981 indicated that 33% of the public was in favor of expanding the number of nuclear power stations in the U.K., while 53% were opposed. Recent European Community surveys provide a more complete picture (Commission of the European Communities, 1982). Averaged over the ten member states, 38% of the public favors further expansion of the nuclear industry, with 37% opposing further development. In 1978 these percentages were 44 and 36 respectively. Averaged over the ten member states the nuclear waste issue was most frequently mentioned as a possible danger of nuclear energy. The second most frequently mentioned issue concerned reactor safety (radioactive emissions from power stations). The above mentioned report to the European Commission concluded that the public is most worried about the storage of radioactive wastes (p. 52).

One of the conclusions of survey studies on public acceptability of nuclear energy is that opinions concerning nuclear waste management must be seen in the wider context of more general attitudes toward nuclear energy. The two matters have become linked in public discussions (see, e.g., National Research Council, 1982; Van der Pligt, 1985). Nealey, Melber and Rankin (1983) report that since 1976 over twice as many people have indicated that they oppose nuclear power because of waste-management problems than was previously the case. Furthermore, when nuclear-power problems were directly compared, waste management was still believed to be a bigger problem than reactor safety. The relationship between the two issues is also apparent in the context of *local* opinions about nuclear developments such as nuclear power plants and waste facilities.

4. LOCAL OPINION ABOUT NUCLEAR FACILITIES

Results of public opinion surveys in both the U.S.A. and Europe show that people are less willing to approve construction of new nuclear facilities in their neighbourhood then to approve the construction of these facilities in general. For instance, support for local nuclear power plants has been in decline since the mid-70s. In the U.S.A.

support decreased from 47% in 1977 to 28% in 1980 (Rankin et al., 1981). A recent study, conducted in three small communities in the South West of England confronted with the possible building of a nuclear power station showed a considerable majority (75%) opposing the plans (Van der Pligt, Eiser and Spears, 1986a). Similar findings are obtained when localities are confronted with hazardous waste facilities. For instance, Cook (1983) found that 74% of the local population in Steele Creek (Charlotte, NC) opposed a proposed waste facility. Similar findings were obtained by Bachrach and Zautra (1985). Other findings indicate that the public is equally opposed to toxic chemical disposal facilities as to nuclear waste facilities; both are accompanied by extremely unfavorable local reaction (Lindell and Earle, 1983).

A number of surveys have either compared level of acceptance of a nuclear facility amongst people who live near one with that of people who do not, or they have monitored local opinion in a locality where the possibility of a nuclear power facility being constructed gradually becomes a reality. Overall, there seems to be mixed support for the idea that familiarity leads to greater acceptance of a nuclear facility in one's community. Melber, Nealy, Hammersla and Rankin (1977) mention eight studies which followed local acceptance of a nuclear power plant as it was being constructed. Only in two of these a significant increase in acceptance over time was found, and one locality showed a significant increase in the level of opposition. Hughey, Lounsbury, Sundstrom and Mattingly (1983) found large negative changes in attitude toward a nuclear facility while being constructed. The weights given to the various aspects of the operation of the facility remained essentially stable over a 5-year period, but people had much lower expectations about potential positive outcomes in the later stages of construction. Results are equally mixed concerning the relationship between living near a nuclear power plant and acceptance of nuclear energy in general (see Thomas and Baillie, 1982). Van der Pligt, Eiser and Spears (1986b) conducted a study showing marginally more favorable attitudes toward nuclear energy in general around Hinkley Point (the site of two existing nuclear power stations in the South-West of England) than in three small local communities that were selected as possible future sites. Other research on local attitudes did not support the notion that familiarity leads to more favorable attitudes (e.g., Warren, 1981). The possible effects of familiarity upon public acceptance have mostly been studies in the context of local nuclear power plants. The relatively short history of nuclear waste management has not been accompanied by similar studies on public acceptance as a function of familiarity.

5. LAY RATIONALITY VERSUS IRRATIONAL FEAR

Two major explanations have been put forward for the recent increases in public opposition. One emphasizes the *emotional, irrational* aspects of human relations to nuclear issues. This view is most popular in psychiatric approaches. For instance, Lifton (1979) argues that the imagery of nuclear weapons has had a profound effect on the human subconscious, and that people's fear for nuclear power issues is an extension of their fear of nuclear weapons. Dupont (1981) characterizes this fear as phobic and believes that this "ultimate irrational fear" is exacerbated by the media's focus on fear in their coverage of the nuclear debate.

A second approach generally assumes that *rational* processes underlie people's decisions about acceptance of nuclear facilities. In this tradition, public concern about nuclear facilities is based on integrating information and relating it to their own subjective values in order to come to an overall judgment. In other words, public opposition is not caused by subliminal fears or phobias but based on everyday inferential strategies.

Explanations that stress the importance of emotial, irrational aspects in opinion formation are usually favored by those supporting nuclear power, with the restriction that these subliminal drives are seen as applicable only to those opposing nuclear energy. It seems unlikely, however, that the antinuclear side of the debate should have a monopoly on subconscious motivations. Furthermore, both sides of the debate rely upon factual arguments and both use legal and scientific experts to strengthen their case. These arguments refer to a wide variety of aspects (economic, political, environmental, and public health). Strictly speaking all these rational arguments could still be traced to subconcious elements, making it rather difficult to test the relative validity of the two approaches.

Mitchell (1984) attempted to compare the two approaches and concluded that the 'lay rationality' approach had greater explanatory power. The major advantage of approaches that assume a rational public is that they provide an explanatory framework that can incorporate a variety of aspects that are seen as relevant. Approaches emphasizing the emotional, uninformed character of public reactions tend to underestimate the relevance of economic, political and environmental arguments by reducing them to manifestations of irrational fear. Although 'lay rationality' approaches have tended to underestimate the role of fear, recent developments focus on decision making models that incorporate elements such as reactions to stress, and coping processes.

In the present chapter, therefore, we will focus on approaches that view people's reactions to the nuclear issue as rational. Within this approach one has attempted to relate people's fears to judgmental and inferential processes that determine people's evaluation of nuclear risks. This issue of risk perception has been extensively studied in recent years, partly with the aim of helping to formulate policy decisions on risk regulation and risk-bearing technologies.

6. RISK PERCEPTION

Although the experts' assessment of the risks of nuclear facilities indicates that these are no greater, and perhaps substantially less than, those of other generally accepted technologies, the public's distrust of nuclear energy is substantial. Opinion polls consistently report qualms about the release of radioactivity, potential catastrophic accidents, and the disposal of nuclear waste. Both operational hazards and possible adverse environmental impacts are seen as major risks of nuclear energy. Initially, attemps to understand people's reactions to the risks of nuclear facilities focused on the contrast between expert judgement and lay people's intuitive assessment of risks. The experts' risk assessments were regarded as objective and quantifiable and public fears were interpreted as biased and irrational. Early work within the tradition of classical decision theory focused on normative models of decision making (e.g., Keeney and Raiffa, 1976), the assumption being that the technical definition of risk (expected losses) is the only tenable approach and that taking the mean losses over time provides a correct way to infer risks from past experience. In this context, the study of cognitive biases and heuristics was expected to help educate the public about risks.

Public disagreement among scientists over the risks of nuclear energy, however, led to the realization that experts' assessments were less 'objective' than previously assumed, and that experts also disagreed about the acceptability of risks. As a consequence research paid more attention to possible factors influencing the perceived acceptability and attempted to develop a framework with which to explain public reactions to technological risks such as those associated with nuclear energy.

A number of studies have revealed that nuclear power, as compared with other technologies, elicits an extraordinary level of concern, particularly because of the characteristics of the hazards that it poses (see, e.g., Fischhoff, Slovic, Lichtenstein, Read and Combs, 1978; Fischhoff, Lichtenstein, Slovic, Derby and Keeney, 1981; Vlek and Stallen, 1981). Most prominent among these are the potentially catastrophic and involuntary nature of possible accidents, and the fact that it is an unknown hazard. Compared to other technologies nuclear energy emerges as the most extreme in terms of the size and seriousness of a potential accident.

The public's concept of risk, therefore, seems to be heavily influenced by qualitative risk characteristics and the catastrophic nature of conceivable accidents. These qualitive factors play a more important role than the assumed probability of the possible negative consequences. The concept of risk, however, does not embrace all the relevant terms of public acceptance. The public's perceptions of risk are built on values, attitudes and sets of attributes which need not be similar to the representations of the experts and policy makers.

7. BELIEFS AND VALUES

Attempts to analyse the structure of people's attitudes toward nuclear energy are usually based on expectancy-value models of attitude formation which broadly assume that the more a person believes the attitude object has good rather than bad attributes or consequences, the more favorable his or her attitude tends to be. Most of the work in this area is based on the model of attitude formation proposed by Fishbein and his collegues (Fishbein, 1963; Fishbein and Hunter, 1964), which analyses attitudes in relation to the anticipated consequences accompanying the attitude object. Results of these studies show that individual attitudes are based upon perceptions of a limited number of potential negative and positive aspects of nuclear energy (e.g., Otway and Fishbein, 1976; Sundstrom, Lounsbury, Schuller, Fowler and Mattingly, 1977; Sundstrom, DeVault and Peelle, 1981).

A further conclusion of this research is that separate dimensions of the issue of nuclear energy appear differentially salient for different attitude groups. Otway, Maurer and Thomas (1978) report the results of a factor analysis on thirty-nine belief statements about nuclear energy. Their results pointed at a number of dimensions underlying the way people think about nuclear energy. These can be summarized as follows:

(a) beliefs about economic benefits,
(b) beliefs about environmental and physical hazards due to routine low-level radiation, and possible accidents,
(c) beliefs about the socio-political implications of nuclear power (e.g., retrictions on civil liberties, increased security measures), and
(d) beliefs about psychological risks (fear, stress, etc.).

In the same study subgroups of the fifty most pro- and fifty most anti-nuclear respondents were compared in order to determine the contribution of each of the four factors to respondents' overall attitudes. For the pro-nuclear group, the economic and technical benefits factor made the most important contribution, whereas for the anti-nuclear group, the risk factors were more important.

Woo and Castore (1980) also found that nuclear proponents attached greater value to potential economic benefits, while the nuclear opponents were more concerned with potential health and safety issues. Results obtained by Eiser and Van der Pligt (1979), Van der Pligt, Van der Linden and Ester (1982), and Van der Pligt et al. (1986a) provide further support for the view that individuals with opposing attitudes tend to see different aspects of nuclear energy as salient, and hence will disagree not only over the likelihood of the various consequences but also over their importance. In other words, each group has its own reasons for holding a particular attitude; the proponents stressing the importance of economic benefits, while the opponents attach greater value to environmental and public health aspects. An important finding of these studies was that the overall attitude of respondents was more closely related to ratings of - in their view - important aspects than to their ratings of subjectively less important aspects. Thus, a consideration of both the perception of the various consequences and the subjective importance or salience provides a more complete picture than could be obtained from a consideration of either factor alone.

The above studies suggest that the attitudinal differences apparent in controversies of this kind require a conception of attitudes and decision making that takes account of the fact that different aspects of the issue will be salient to the different sides of the debate, and that such differences in salience may be at least as clear-cut and informative as differences in the likelihood and evaluation of the various potential consequences. As argued elsewhere (Van der Pligt and Eiser, 1984) the finding that separate dimensions of the issue appear differentially salient (both subjectively and in their contribution to the prediction of overall attitude) for the different attitude groups, has important practical implications for theories of attitude and our understanding of why people hold different attitudes toward nuclear energy.

A further study (Van der Pligt et al., 1986a) focused on *local* attitudes toward the building of a nuclear facility. Results underlined the importance of including both beliefs and salience in one's conception of attitude. For instance, even though the attitude groups (pro vs. anti) showed relatively minor differences in there evaluation of the effects of potential employment opportunities in the locality, a majority of the pros found this aspect important, while only a small minority of the antis regarded this aspect as being of importance. Results of this study also showed that the major differences between the attitude groups concern the less tangible, more long-term nature of te potential negative outcomes.

Findings further suggested that the perception of the psychological risks is the prime determinant of attitude. In this study subjects were presented with two sets of fifteen potential consequences. Overall, differences in perceived importance of the various consequences showed that the antis stressed the importance of the potential risks for public health and the environment. A stepwise multiple regression analysis, with participants' attitude toward building a nuclear power station in their neighborhood as a dependent variable and their ratings on all thirty possible consequences as independent variables, provided strong evidence for the importance of psychological risks.

The most striking aspect of the analysis was the predictive power of the aspect 'personal peace of mind'. This issue dominated the analysis, and expectations about one's peace of mind correlated 0.79 with attitudes toward the nuclear facility. The remaining aspects (economic, socio-political) made only marginal contributions to the prediction of people's attitudes toward the building and operation of the nuclear facility.

In summary, opponents and proponents of nuclear facilities have very *different* views on the possible consequences. This applies to both the general issue of nuclear energy and to the building of a nuclear facility in one's locality. The most significant difference, however, concerns the perception of psychological risks (anxiety, stress). This factor becomes more important when people are (or will be) more directly exposed to the risks, for instance when their locality is selected as a possible site for a nuclear facility. The importance of stress-related variables has been shown in a wide variety of studies of local reactions to the risks of nuclear facilities (e.g., Baum, Fleming and Singer, 1982), soil contamination (De Boer, 1986) and industrial hazards (e.g. Stallen and Tomas, 1988).

These findings also suggest that the different perceptions of the possible consequences of further expansion of the nuclear industry are related to more general values. Public acceptance of nuclear facilities is not simply a matter of perceptions of risks but is also related to more generic issues such as the value of economic growth, high technology and centralization (Van der Pligt, 1985). The issue of nuclear energy is embedded in a much wider moral and political domain.

Both sides of the issue believe that the debate is about more than just technical concerns. It needs to be added, however, that more general aspects such as economic growth, and especially, the social impact of technological developments and their political consequences figure most prominently in assessments of those active in the debate (see, e.g., Eiser and Van der Pligt, 1979). Some of these aspects (e.g., political consequences, security measures) seem less relevant to the general public; others (e.g., the value of economic growth) do play a role in the general public's assessment of the issue. At a local level such as in siting disputes, safety aspects tend to dominate public perception. Not surprisingly, the antinuclear movement generally regards the safety issue as their natural point of contact with the public (see, e.g., Gyorgy, 1979; Mitchell, 1984).

Douglas and Wildavsky (1982) stress the importance of social and moral evaluations of specific hazards or hazard distributions. This leads us to the issue of equity. This issue is of crucial importance in siting efforts of waste facilities. As compared to nuclear power stations, the economic benefits (e.g., employment opportunities) of waste facilities are marginal. This leads to situations where some members of society are exposed to risks without receiving sufficient compensation for these potential costs. In the next section, the equity issue and other conclusions based on research findings presented in earlier parts of this chapter will be discussed in the context of siting radioactive wastes.

8. SITING POLICY, EQUITY AND PUBLIC PARTICIPATION

Kasperson (1985) addresses the widely held opinion that inequity is the crucial problem for hazardous waste facility siting. Once a locality is shortlisted, the community objects to being the dumping ground for waste from elsewhere. Furthermore, the affected locality is likely to oppose the facility because the benefits will flow to others (waste-generating industries, the general public and the owners of the facility) while the risks will be concentrated locally.

This geographical dissociation of costs (risks) and benefits can not easily be balanced. As discussed earlier in this chapter, the perceived risks and the qualitative nature of these risks associated with considerable fear, dominate local acceptance of nuclear facilities. These aspects simply overwhelm any prospect of restoring the original conditions through the enlargements of benefits.

This view is supported by recent experiences where the relevant authorities attempted to increase local acceptance of a nuclear waste facility by providing compensation to the locality. The purposes of compensation (e.g., financial, economic) are to change local motivation to oppose the facility, increase the efficiency of facility planning because costs and benefits are better accounted for, and finally, to promote negotiation, as opposed to confrontation, in reaching siting decisions (see, e.g., O'Hare, Bacow and Sanderson, 1983). Because the environmental and health risks so dominate public reactions, the prospect of compensation does not effectively lower the degree of local resistance or engender a propensity among local residents to "trade off" concerns. Moreover, compensation tends to be viewed as a "bribe", exacerbating the equity issue and increasing suspicion and distrust of the relevant authorities (see, e.g., Kasperson, 1985; Kunreuther, Linnerooth and Vaupel, 1984).

A number of states have used compensation strategies in cases of siting hazardous non-radioactive wastes. Massachusetts adopted an approach based upon the work of O'Hare et al. (1983). Kasperson

(1985) lists the following key ingredients. The two primary parties in the bargaining process are the host community and the developer, who are required to reach a settlement (via negotiation or arbitration) including mitigation and compensation to the host community. The community has only a limited basis on which to refuse the construction of the facility, and impasses between developer and the affected community are submitted to an arbitrator.

Three assumptions are of particular interest in the context of the approach discussed by O'Hare et al. (1983). First, voluntary consent is assumed to be achievable through sufficient provision of incentives and through direct bilateral negotiations between the community and the developer. Second, the long-term consequences of the facility can be defined with sufficient precision to formulate an acceptable and appropriate compensation package (e.g., economic benefits, financial help, provision of other facilities that benefit the community). Finally, it is assumed that both the developer and/or the relevant regulatory authorities can rely upon sufficient social trust to reassure local fears and to prevent a conflict-laden decision making process.

The last assumption seems rather optimistic given the erosion of public confidence in a wide variety of major social institutions (Public Opinion, 1980). Portney (1983) conducted a survey in Massachusetts indicating that lack of trust in both the management of companies that operate the facilities, and the government regulators who monitor whether proper procedures are followed, is a significant source of concern for the community. Similar remarks can be made about the second assumption. The literature on social and environmental impact assessment acknowledges the limited ability to identify and/or measure the possible consequences of the building and operation of a siting facility.

A further problem concerns the difficulties in translating health and environmental risks into compensating economic measures or financial benefits. It is not surprising, therefore, that siting efforts based on compensation measures have had mixed success. This lack of success provides further support for the earlier presented view that local opinion near nuclear facilities tends to be dominated by perceived environmental and health risks. This is especially the case in the early stages of planning procedures. Economic benefits and compensation are of secondary importance to residents of communities confronted with the possible location of a nuclear waste facility in their neighborhood.

The above discussion suggests that public acceptance may well depend more on the characteristics of the *process* that allocates risks than on the relationship between risk and benefits. Compensation does not seem a sufficient condition to increase public acceptance of nuclear facilities. This lack of public acceptance, could well be related to the limited *public participation* in decision making processes concerning siting nuclear waste and to the *communication* between the experts (and relevant authorities) and the public.

Since safety-related issues play a crucial role in public acceptance of waste facilities, it seems necessary to improve the relationship between the expert and the lay public. For the experts this poses an important challenge: to recognize the limitations and fallability of risk assessments, and to be aware of the fact that important qualitative aspects of risks influence the reponses of lay people. For lay people it seems necessary to accept the necessity to be better informed and to be aware of the influence of these qualitative aspects.

The high level of concern and involvement of residents of communities confronted with the possible construction of a nuclear facility poses a futher challenge to risk communication. Earle and Cvetkovich (1986) point at the importance of developing a common framework appropriate to the particular hazard and acceptable to all involved parties. Development of this framework is a necessary condition for risk communication (e.g., information about the specific health and environmental risks, information about possible ways to reduce the effects of the hazard) to be succesful. Experience has shown that communication about risks is extremely difficult and often frustrating to those involved. Government officials and experts frequently complain about the lack of understanding of lay people and the distorted and biased media coverage. Individual citizens, on the other hand, often perceive a lack of interest in their concerns, and a reluctance to allow them to participate in decisions that intimately affect their lives. Recent experiences in the Netherlands are in accordance with the above view (see, e.g, Brouwer, this volume). Attempts to inform the public in a number of communities shortlisted as possible sites for radioactive waste were largely unsuccesful and were frequently accompanied by demonstrations of the local residents. It is beyond the scope of this chapter to discuss the many relevant aspects of risk communication. Cvetkovich, Earle and Vlek (this volume) address this issue in more detail.

Communication will play an important role in increased public participation in decisions about siting radioactive waste. Increased participation could improve the fairness of the risk allocation process and increase the degree of trust in the authorities responsible for the siting problem. Experience with approaches that incorporate public participation is limited. This approach entails more than a public relations task. It demands more openness of a traditionally closed industry, a different distribution of knowledge and expertise, and substantially improved communication between experts and the public. A necessary condition for the latter is more mutual understanding and respect for the concerns and representations of the parties involved in the siting problem.

Psychological decision theory could help clarifying the many differences of opinion about facts and values that play a role in siting disputes (see, e.g., Chen and Mathes, this volume). In the face of growing evidence that the current approaches are failing, increased public participation could well be the only way to reach acceptable solutions and increase the dramatically eroded trust and credibility of the nuclear power industry.

REFERENCES

Bachrach, K.M. and Zautra, A.J. (1985). Coping with a community stressor: The threat of a hazardous waste facility. *Journal of Health and Social Behavior, 26,* 127-141.

Baum, A., Fleming, R. and Singer, J.E. (1982). Stress at Three Mile Island: Applying social impact analysis. In: *Applied Social Psychology Annual.* Beverly Hills: Sage.

Brouwer, H.C.G.M. (this volume). Current radioactive waste management policy in the Netherlands.

Chen, K. and Mathes J.C. (this volume). Value oriented social decision analysis: a communication tool for public decision making on technological projects.

Commission of the European Communities (1982). *Public opinion in the European Community* (Report No. XVII/202/83-E). Brussels, Belgium: Commission of the European Communities.

Cook, J.R. (1983). Citizen response in a neighborhood under threat. *American Journal of Community Psychology, 11,* 459-471.

Cvetkovich, G., Vlek, Ch. and Earle, T.C. (this volume). Designing technological hazard information programs: towards a model of risk-adaptive decision making.

De Boer, J. (1986). Community response to soil contamination: Risk and uncertainty. In: J.W. Assink and W.J. van der Brink (Eds.), *Contaminated soil*. Dordrecht: Martinus Nijhoff Publishers.

Douglas, M. and Wildavsky, A. (1982). *Risk and culture*. Berkely, CA: University of California Press.

Dupont, R. (1981). The nuclear power phobia. *Business week, september 7*, 14-16.

Earle, T.C. and Cvetkovich, G. (1986). Failure and success in public risk communication. In: The Air Pollution Control Association (Eds.), *Avoiding and Managing Environmental Damage from Major Industrial Accidents*. Pittsburgh (PA): Air Pollution Control Association.

Eiser, J.R. and Van der Pligt, J. (1979). Beliefs and values in the nuclear debate. *Journal of Applied Social Psychology, 9*, 524-536.

Fischhoff, B., Lichtenstein, S., Slovic, P., Derby, S.L. and Keeney, R.L. (1981). *Acceptable risk*. Cambridge: Cambridge University Press.

Fischhoff, B., Slovic, P., Lichtenstein, S., Read, S. and Combs, B. (1978). How safe is safe enough: A psychometric study of attitudes toward technological risks and benefits. *Policy Sciences, 8*, 127-152.

Fishbein, M. (1963). An investigation of the relationship between beliefs about an object and the attitude towards that object. *Human Relations, 16*, 233-240.

Fishbein, M. and Hunter, R. (1964). Summation versus balance in attitude organization and change. *Journal of Abnormal and Social Psychology, 69*, 505-510.

Gyorgy, A. (1979). *No nukes: Everyone's guide to nuclear power*. Boston MA: South End Press.

Hughey, J.B., Lounsbury, J.W., Sundstrom, E. and Mattingly, T.J. (1983). Changing expectations: A longitudinal study of community attitudes toward a nuclear power plant. *American Journal of Community Psychology, 11*, 655-672.

Kasperson, R.E. (1985). *Rethinking the siting of hazardous waste facilities*. Paper presented at the Conference on Transport, Storage, and Disposal of Hazardous Materials. IIASA, Wenen, Oostenrijk; juli.

Kasperson, R.E., Berk, G., Pijawka, D., Sharaf, A.B. and Wood, J. (1980). Public opposition to nuclear energy: Retrospect and prospect. *Science, Technology and Human Values, 5*, 11-23.

Keeney, R.L. and Raiffa, H. (1976). *Decisions with multiple objectives: Preferences and value tradeoffs*. New York, Wiley.

Kunreuther, H., Linnerooth, J. and Vaupel, J.W. (1984). A decision-process perspective on risk and policy analysis. *Management Science, 30 (april),* 475-485.

Lifton, R.J. (1979). *The broken connection: on death and the continuity of life.* New York: Simon and Schuster.

Lindell, M.K. and Earle, T.C. (1983). How close is close enough: Public perceptions of the risks of industrial facilities. *Risk Analysis, 3,* 245-253.

Melber, B.D., Nealey, S.M., Hammersla, J. and Rankin, W.L. (1977). *Nuclear power and the public: analysis of collected survey research.* Batelle Memorial Institute, Human Affairs Research Centres, Seattle, Washington.

Mitchell, R.C. (1984). Rationality and irrationality in the public's perception of nuclear power. In: W.R. Freudenburg and E.A. Rosa (Eds.), *Public reactions to nuclear power: Are there critical masses?* (pp. 137-179). Boulder: Westview Press.

National Research Council (1982). Radioactive waste management. In: *Outlook for science and technology.* San Francisco, W.H. Freeman.

Nealy, S.M., Melber, B.D. and Rankin, W.L. (1983). *Public opinion and nuclear energy.* Lexington, M.A.: Lexington Books.

O'Hare, M., Bacow, L. and Sanderson, D. (1983). *Facility siting and public opposition.* New York: Van Nostrand.

Otway, H.J. and Fishbein, M. (1976). *The determinants of attitude formation: An application to nuclear power.* (Research Memorandum RM-76-80). Laxenburg, Austria: International Institute for Applied Systems Analysis.

Otway, H.J., Maurer, D. and Thomas, K. (1978). Nuclear power: The question of public acceptance. *Futures, 10,* 109-118.

Portney, K.E. (1983). *Citizen attitudes toward hazardous waste facility siting: Public opinion in five Massachusetts communities.* Medford, MA: Tufts University (Center for Citizenship Public Affairs).

Public Opinion (1980). Surveys by Louis Harris and Associates summarized in *Public Opinion, 3(2),* page 26.

Rankin, W.L., Melber, B.D., Overcast, T.D. and Nealy, S.M. (1981). *Nuclear power and the public: An update of collected survey research on nuclear power.* (PNL-4048). Batelle Human Affairs Research Centres, Seattle, Washington.

Stallen, P.J.M. and Tomas, A. (1988). Public concern about industrial hazards. *Risk Analysis, 8,* 237-245

Sundstrom, E., DeVault, R.C. and Peelle, E. (1981). Acceptance of a nuclear power plant: applications of the expectancy-value model. In: A. Baum and J.E. Singer (Eds.), *Advances in environmental psychology, Volume 3* (pp. 171-189). Hillsdale, N.J.: Lawrence

Erlbaum.

Sundstrom, E., Lounsbury, J.W., Schuller, C.R., Fowler, J.R. and Mattingly, T.J. Jr. (1977). Community attitudes toward a proposed nuclear power generating facility as a function of expected outcomes. *Journal of Community Psychology, 5,* 199-208.

Thomas, K. and Baillie, A. (1982). *Public attitudes to the risks, costs and benefits of nuclear power.* Paper presented at a joint SERC/SSRC seminar on research into nuclear power development policies in Britain, June.

U.S. Environmental Protection Agency (1979). *Draft environmental impact statement for subtitle C.* Washington: EPA.

Van der Pligt, J. (1985). Public attitudes to nuclear energy: Salience and anxiety. *Journal of Environmental Psychology, 5,* 87-97.

Van der Pligt, J. and Eiser, J.R. (1984). Dimensional salience, judgment and attitudes. In: J.R. Eiser (Ed.), *Attitudinal judgment* (pp. 161-177). New York, Springer.

Van der Pligt, J., Eiser, J.R. and Spears, R. (1986a). Construction of a nuclear power station in one's locality: Attitudes and salience. *Basic and Applied Social Psychology, 7,* 1-15.

Van der Pligt, J., Eiser, J.R. and Spears, R. (1986b). Attitudes toward nuclear energy: Familiarity and salience. *Environment and Behavior, 18,* 75-93.

Van der Pligt, J., Van der Linden, J. and Ester, P. (1982). Attitudes to nuclear energy: Beliefs, values and false consensus. *Journal of Environmental Psychology, 2,* 221-331.

Vlek, Ch. and Stallen, P.J.M. (1981). Judging risks and benefits in the small and in the large. *Organizational Behavior and Human Performance, 28,* 235-271.

Warren, D.S. (1981). *Local attitudes to the proposed Sizewell 'B' nuclear reactor.* Report RE 19. Food and Energy Research Centre, October 1981.

Welch, M.J. (1985). Nuclear-waste disposal reaches critical stage. *New York Times (March 20):* A26.

Woo, T.O. and Castore, C.H. (1980). Expectancy-value and selective exposure determinants of attitudes toward a nuclear power plant. *Journal of Applied Social Psychology, 10,* 224-234.

Yankelovich, D. and Kaagan, L. (1981). *The American public looks at nuclear: Vaguely in favor, clearly worried.* Pittsburg, PA: Aluminum Company of America.

DESIGNING TECHNOLOGICAL HAZARD INFORMATION PROGRAMS: TOWARDS A MODEL OF RISK-ADAPTIVE DECISION MAKING

GEORGE CVETKOVICH, CHARLES VLEK AND TIMOTHY C. EARLE

Department of Psychology
Western Washington University, Bellingham

Department of Psychology
University of Groningen

1. INTRODUCTION

In the late Summer of 1985 the Dutch Ministry of Economic Affairs organized a series of informational meetings throughout the country on the siting of new nuclear power plants. The meetings were characterized by an extremely hostile response to the government's pro-nuclear position. Most meetings were seriously disrupted by continuous demonstrations of environmental protection groups. Asked to explain the reaction of his fellow citizens, the Ministry's director-general of Energy stated that it was due to a lack of information about the need for, the economic benefits of, and the nature of nuclear power.

This statement was particularly remarkable because a costly, two and one-half year government-supported "wide societal discussion" on the possible expansion of nuclear power, involving large portions of the population had been concluded in 1984 (see Vlek, 1986). During the nation-wide debate varied information on the possible expansion of nuclear power had been analyzed and widely disseminated through public meetings and the media. The outcome of this public discussion was controversial; it showed a public majority against nuclear power expansion to be in opposition against a powerful industrial-economic minority which favored it. The Dutch Minister of Economic Affairs had discovered, as have many other administrators of technology development programs, that communicating about technology may not always be conclusive, even after extended effort.

Ch. Vlek and G. Cvetkovich (eds.), Social Decision Methodology for Technological Projects, 253–275.

Despite the difficulties, there is a growing effort to communicate to the public about hazardous technologies. At about the same time as the Dutch nuclear debate William Ruckelshaus (1983), departing head of the U.S. Environmental Protection Agency, remarked:

"To effectively manage ... risk, we must seek new ways to involve the public in the decision making process ... They need to be informed if their participation is to be meaningful ... We must search for ways to describe risk as clearly as possible, tell people what the known or suspected health problems are, and help them compare that risk to those with which they are more familiar."

General values about citizens' rights and needs to know about the risks they are exposed to, are developing, while the duties of industrial companies and government agencies to make hazard information public, are being emphasized (Baram, 1984).

In the wake of such disasters as have occurred in Flixborough (U.K.), Seveso (Italy), Harrisburg (Pennsylvania), Bhopal (India), Mexico City (the LPG fire), Chernobyl (U.S.S.R.) and Basel (the Rhine pollution from the Sandoz fire), the Commission of the European Communities (1982) is actively seeking to implement article 8 of its 1982 "post-Seveso directive", which says:

"Member states shall ensure that persons liable to be affected by a major accident originating in a notified industrial activity ... are informed in an appropriate manner of the safety measures and the correct behaviour to adopt in the event of an accident."

Thus, both legal and non-legal moves have been made to ensure that effective hazard communication will occur. Regrettably, however, hopes for the positive effects of communication about hazardous technologies often far outstrip actual outcomes (see, e.g., Brouwer, this volume). This disappointing state of affairs may be ascribed to lacking insights into the informational needs and wishes of public audiences, their likely manner of information processing, and the behavioral decisions that are, or may be, perceived as 'correct' in the event of an accident or an impending disaster.

It is the purpose of this chapter to review various theoretical bases for designing public hazard information programs, and to discuss their assumptions and implications. Later in the chapter we will present a framework for understanding and modelling the public's 'risk-adaptive decision making'. Thus the chapter is concentrated upon the proper content of hazard information programs rather than upon the peculiarities and effects of mass media communication of risk information (see Peltu, 1985, for a pertinent review).

Generally speaking the goal of hazard communication is to improve the public's ability to make appropriate decisions in view of technological hazards (and benefits). But, what does "appropriate decisions" mean and how could hazard communicators best achieve this end? Answers to these questions are dependent on an understanding of the components of the process of communication. Our conceptualization of the communication process is presented in Section 2 of this chapter.

In Section 3, three prevalent models of communicating hazard information to public audiences are reviewed. Each model includes assumptions about the goals of hazard communication, the needs of the audience, and the way in which audience members make decisions and react to communicated information. We conclude that while existing models offer many useful suggestions to the hazard communicator, for different reasons each is insufficient and may lead to important miscommunications.

With this review of the current state of the field as background, Section 4 presents an overview of the fundamentals of a decision-theoretic approach to hazard communication. This model of 'risk-adaptive decision making' is based on a response-definition of risk and includes an assessment of desirable and undesirable scenarios. The final section of the chapter discusses some of the implications and challenges to the hazard communicator presented by the decision-theoretic model.

2. COMPONENTS OF THE COMMUNICATION PROCESS

As shown in Figure 1, communication begins with the communicator's view, or image, of reality (Step 1). This contains both an image of the hazardous technology, including an assessment of its risks, as well as the conclusion that there is a need for communication with specified audiences, if the communicator is to accomplish his or her goals. Thus the communicator's view of reality determines both that communication will occur and also what the goals of that communication should be.

The second component of communication is the image of the audience held by the communicator   (Step 2). This   image contains   assumptions

Figure 1: Component processes of hazard communication.

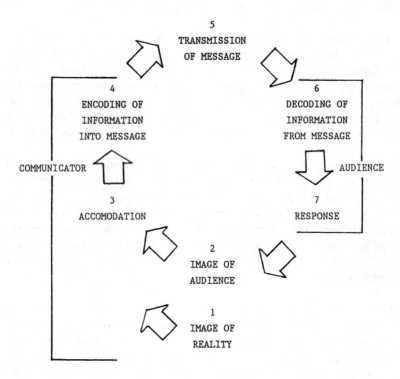

about   knowledge   and   abilities   (e.g.,   "the   audience   doesn't   have
technical   training"),   motivation   for   personal   behavior   concerning   a
technology,   including   information   seeking   (e.g.,   "the   audience   is
uninterested   in   hazard   information"),   personal   needs   (e.g.,   "the
audience   should   know   more   about   safety   precautions"),   the   process   of
assessing   credibility   of   messages   (e.g.,   "the   audience   doesn't   trust
information   from   the   Environmental   Protection   Agency"),   and   the
nature   of   judgment   and   decision   making   including   how   the   audience
defines   risk   (e.g.,   "the   audience   only   uses   severity   of   outcome,   not
probability   of   occurrence   in   judging   the   riskiness   of   a   technology").

The   images   of   Steps   1   and   2   are   combined   through   an   interactive
process   of   accommodation   (Step   3)   whereby   the   communicator   adapts
the   production,   organization,   argument   and   mode   of   presentation   of   a
message   to   fit   the   perceived   characteristics   of   the   audience,   the

communicated information, and the situation in which the message occurs (Cvetkovich, 1978).

Next the message is constructed by using a symbol system (e.g., language, pictures, graphs, etc.) to encode information from the communicator's representation of reality, as accommodated on the basis of his or her images of the audience (Step 4), and it is transmitted to the audience (Step 5). The audience likewise uses a symbol system (not necessarily the same) to decode the information contained in the message (Step 6). The particular match between the symbol systems used by the audience and that used by the communicator is determined by characteristics of the audience such as their background training and experience, their socio-political views, and their resulting motives and needs.

If the communicator has taken the step of gathering information about the audience's responses (Step 7), changes may occur in the communicator's images of the audience. If this is so, the process of hazard communication becomes a case of two-way communication, with the possibility of iterative improvements occurring in the communicator's ability to accurately convey information.

From the perspective of Figure 1, effective communication is defined as the extent to which the audience decodes from the message, and accepts as credible, information that matches the information which the communicator intended to transmit. Thus, effectiveness of (hazard) communication is largely determined by the adequacy of the accommodations made by the communicator to his or her view of reality. It is the premise of this chapter that communication effectiveness can be improved if hazard communicators assume a decision-theoretic perspective to guide this accommodation process. Before going on to specify, in Section 4, the nature of this framework, we will review three current models of public hazard communication.

3. HAZARD COMMUNICATION MODELS AND THEIR AUDIENCE VIEWS

Current approaches towards providing hazard information to the public largely follow from three different models of the public's judgment and decision making processes: (a) the Factual Information Model, (b) the Personal Gain/Loss Model, and (c) the Values Model.

3.1 The Factual Information Model

This perspective sees the problems of communication as consisting of questions about whether the communicator is able and willing to provide enough appropriate facts, the (in)ability of the audience to understand the facts presented, and whether, in the view of the audience, a persuasive argument based on the evidence is made. Hazard messages may not be understood because they are couched in technical terms and the information is based on methods of analysis not understood by the public.

For example, people sometimes have difficulty understanding probabilities and how base-rate information applies to their particular cases (see Covello, Von Winterfeldt and Slovic, 1986; Slovic, 1986). This suggests that efforts should be directed towards public instruction in the interpretation of probabilities or the development of alternative methods of conveying hazard information.

Many people also have an inadequate understanding of the decision processes used to develop and site hazardous technologies and of the nature of the methods underlying these decisions, such as probabilistic risk analysis. Thus they may misinterpret the results of risk analysis or fail to adequately apply them. This suggests that hazard communicators need to take greater care in educating the public about technology assessment methods.

Increasing the public's knowledge about hazards should be an important communication goal. However, the Factual Information Model does not provide an adequate conceptualization for the development of effective communication programs. Communication programs based solely on this model have not been very successful in reaching their designed ends. The Factual Information Model fails because of its faulty assumptions. Briefly these are:

(1) *The assumption that the audience imagines reality in a manner similar to the communicator.*

In recommending that communicators abandon this assumption and make an effort at greater accommodation, Earle and Cvetkovich (1985) argue for a "marketing orientation" to hazard communication. This orientation is characterized by two points: (a) the marketing concept, i.e., success in communication depends on providing information that people want; and (b) the market segmentation concept, i.e., a communicator working with any large audience must make distinctions related to individual and group differences in information needs. If it is assumed that an important motivation for people is to make sense out of the world, information needs related to cognitive functioning should be particularly dominant in most hazard situations.

It follows that if information is to be attended to, remembered and used, it must relate to people's existing mental images, i.e., the modifiable information structures which represent people's understandings of a particular phenomenon and their basic beliefs about it (Cvetkovich, Earle, Schinke, Trimble and Gilchrist, 1986; Cvetkovich and Keren, 1988; Fishbein and Ajzen, 1981). Hazard images consist of organized information concerning the potential risks of a technology as well as relevant information about possible consequences, costs, benefits, and other "facts" in association to values and more general related beliefs.

Mental images of interest to the hazard communicator consist of both those relevant to the workings of particular technologies and those relating to the symbolic or signal value of hazard messages (Cvetkovich and Keren, 1988; Slovic, 1986). Acceptance of a marketing orientation to communication implies the necessity of empirical research which provides the communicator with the information required for appropriate accommodation to various audiences.

(2) *The assumption that the audience uses analytic information processing.*

For the content of a factual message to have an effect the individual must attend to the message, comprehend it, remember the evidence presented and analyze and evaluate the argument. Such "deep cognitive processing" of messages leads to changes in attitudes and other aspects of thinking according to a *central* route of persuasion (Petty and Cacioppo, 1981).

Among the conditions that prevent the deep processing of technological issues are that people may not have the necessary training or abilities to fully analyze the issue, they may lack the time for a detailed consideration of the issue, they may not be motivated to do so because of other interests or a general lack of interest in technological/scientific matters, or they may be "over-motivated" and highly emotional in their reaction because of the great fear aroused by their concern about a technology.

Because one or more of these conditions may apply to many individuals with regard to technological issues it is expected that *peripheral* ('shallow') routes of processing information about hazardous technologies are more frequent than central routes. In a peripheral route, the individual is not involved in a detailed evaluation of the message's information, arguments and recommendations. Rather the individual is motivated to short-term changes by the association of persuasion cues with the message. One study estimates that 60% of risk communicators prefer to use communications geared to a central-route, analytic-thinking decision making style and that this style will miss 75% of any general audience (Johnson and Petcovic, 1986).

(3) *The assumption that hazard "facts" are apolitical.*

What is intended to be a simple communication of facts often becomes an issue of risk assessment (Kasperson, 1986). Adherents to the Factual Information Model tend to view communication with the public as one-sided (from "us" to "them") and simply as a means to reaching an end such as "educating the public", "building support for our project", or "correcting misperceptions".

Citizens, on the other hand, may view the attempted communication as an opportunity for opening discussion and two-way interaction and as a means towards developing new conclusions about tolerable levels of risk and the processes for deciding what they are. As Van der Pligt & Eiser (1984; also Van der Pligt, this volume) have shown, resistance to (or support of) a hazardous technology is not simply a matter of assessed risk based on known facts, but it is associated with salient basic values. To assume that resistance to technological development is only based on ignorance, as for example some nuclear power promoters have done, reflects too narrow an analysis (cf. Hisschemöller and Midden's 'technocratic approach'; this volume).

3.2 The Personal Gain/Loss Model

Underlying communications based on the Personal Gain/Loss model is the assumption that people are motivated by the desire to maximize short-term gains and reduce losses. Hence, communicators should convey information about the utility of potentially hazardous technologies. The Personal Gain/Loss Model differs from the Factual Information Model in that it recommends supplying information about the (personal) utility of outcomes resulting from a technology (as opposed to the general workings of technology and their hazard implications).

Although developers of communication programs do not always draw their implications directly from formal theory, typical use of the utility concept holds conclusions and assumptions similar to those of classic economic theory and its offspring: contemporary analytic decision theory. One difference is that classical economic theory holds that gains and losses are objectively defined. Modern analytic decision theory allows that utilities are subjectively defined. As a foundation of hazard communication programs, however, the utilities and the processes used to derive them are very much presented as objective facts.

Information about probabilities, losses and gains that is included in hazard communications based on the Personal Gain/Loss model is most often derived from risk and risk-benefit analyses. It is strongly implied in communications to the public that this information is objective and more accurate than that possessed by the audience.

The subjectivity of risk assessment, resulting from assumptions, lack of total knowledge about systems, and the failure to sufficiently imagine all the things that could possibly go wrong, is often overlooked, at least in initial communications. Intuitively the audience may recognize, as many practitioners and critics of risk analysis have done explicitly, the flaws in these assumptions. Thus communication of the results of risk analysis may initiate an exchange that soon bogs down in a fruitless conflict focusing on issues concerning the validity of the analysis.

Adherents to the Personal Gain/Loss Model implicitly or explicitly assume a particular process of decision making not too dissimilar from that assumed by the Factual Information Model. Unlike the Factual Information Model, however, this model provides a more general conceptualization and assessment of support or opposition to hazardous technologies. The public is expected to oppose technologies when personal losses may be high or direct personal gains are low. The Personal Gain/Loss Model results in suggestions not only about the content of communication but also about the possible conduct of a program of citizen interaction and participation.

For example, the model might suggest that the utility of a hazardous facility for a local community should be changed through the use of compensation programs that off-set perceived losses (O'Hare, 1977; Kasperson, 1986). Such application of the Personal Gain/Loss model is prone to criticisms that officials are trying to "buy out" the citizens and that they are more concerned about the speedy approval of their plans than they are about valid assessments of risks (see Hisschemöller and Midden's 'market approach'; this volume).

Paradoxically, while this model focuses on possible personal gains and losses, public audiences often find it difficult to apply this information to themselves. The reason for this is that information is provided in the form of "macro- probabilities", frequentistic information about the occurrences of events over large populations. Any single individual may have difficulty in understanding exactly how these probabilities might apply to his or her unique case.

3.3 The Values Model

This model offers an alternative to the assumptions that people's attitudes and actions towards a hazardous technology are based on either their amount of factual knowledge about it or on their personal gain-loss balance resulting from it. Risk judgment and decision making are viewed under this model relative to the public's understanding of the meaning of a technology within a broader context of value judgments. One important consideration is the individual's values concerning judgments of fairness or justice (Tyler, 1986). The Values Model embeds the effect of hazard communication within the matrix of other information that the person is getting and inferring about the wider context of the technology, the society in which (s)he lives, and beliefs about the operations and policies of a particular government administration.

In essence, this perspective focuses on citizens' evaluations of the political process. Research distinguishes between two alternative aspects of judgments of justice - *procedural* justice, the manner in which judgments and decisions in society are made, and *distributive* justice, the way in which actual outcomes such as technological risks and benefits are distributed among different groups in society.

While theories of distributive justice differ from the Personal Gain/Loss Model in placing a greater emphasis on social and ethical judgments, they still assume the pre-eminent importance of outcomes. Theories of procedural justice, in comparison, are concerned with the manner in which hazardous technology decisions are made and the implications these processes have for the protection of public safety and the distribution of risks and benefits among various subgroups (cf. Hisschemöller and Midden's 'distributive justice approach'; this volume).

Several contributors to this volume discuss how consideration of value differences might be incorporated into the social decision making of small face-to-face groups (Bezembinder; Chen and Mathes; Phillips). Integration of the Values Model perspective into communications to large public audiences represents one of the biggest challenges to the hazard communicator. How can observations about the importance of justice (especially procedural justice) be of practical use to the hazard communicator?

Rayner (1986; Rayner and Cantor, 1987) argues that programs should be developed on the basis of known group differences in judgments of justice. A group's social organization and the way in which the individual is typically integrated within his/her group produces sensitivities to different aspects of risk-relevant problems and preferences for different decision strategies (e.g., Douglas and Wildavsky, 1982) including the judged fairness of risk distribution and compensation programs (Thaler, 1985; Törnblom and Johnson, 1985; Wilke, this volume).

For example, individuals influenced by the values of competitive individualistic groups prefer policies in which free market forces are allowed to determine technology planning priorities. In contrast, individuals influenced by stratified bureaucratic groups prefer policies where citizens have entered an assumed social contract with the technology planning agency. For these citizens it is assumed that in exchange for their acceptance of a technology, they can expect that the technology will be operated and controlled in an acceptable manner. In keeping with the market orientation discussed in Section 3.1, this line of reasoning advises that programs be designed relative to group differences (or at least be rationalized in terms "understandable" to the group). The danger of this is that it may leave the impression that the communicator is simply using the words that the audience wants to hear. Effective communication must not only be understandable but must validly reflect procedural realities that are acceptable to the audience.

It is also possible that procedural justice alone will not lead to satisfactory conclusions to conflicts over technologies. The "open" communication policy of the U.S. Environmental Protection Agency's management of emission standard setting for the ASARCO Copper Smelter in Tacoma, Washington, produced the desirable effects of getting large numbers of people personally involved with workshops and hearings, of reducing suspicions about government actions, and of developing a sense of open dialogue (Baird, 1986; Baird, Earle, and Cvetkovich, 1986; Earle and Cvetkovich, 1986).

Regrettably the program was not a success relative to the goals of providing factual information. Many citizens reported being overwhelmed and confused by the technical information and a majority were unable to answer simple questions about emission hazards and the purposes of the EPA actions. This extends Tyler, Rasinski and Griffin's (1986) conclusion that judgments of justice may be independent of other judgments such as personal gain/loss or need for factual information, and it strongly suggests the need for a taxonomy of communication situations. The latter might indicate when each of the issues highlighted by each of the three communication models reviewed is relevant.

3.4 Summary: Hazard communicators' models - a need for improvement

Our review indicates that current models of public hazard communication offer a number of important and useful suggestions on how communicators should accommodate their view of reality to that of their audience. It also indicates that the prevalent models are in many ways inadequate to their task and that improvements are needed. There are aspects of current communicators' models that lead to regularly occurring shortfalls in accommodation, which may result in critical miscommunications.

The differences between communicators and their publics may rest in: (1) the premises underlying risk judgments (e.g., a management perspective versus concern with the social distribution of risk); (2) the unit of analysis assumed (e.g., aggregate versus individual risk); (3) differences in the time perspective taken (e.g., past experience versus future prevention); and (4) the decision making processes assumed to operate.

Because of inadequacies in current models, communicators also often fail to make needed differentiations of subgroups in their audience and do not adjust hazard information so that it is pertinent to local conditions. They may further fail by overlooking real everyday constraints that influence the public's decision making such as connections to an employer who operates a hazardous technology facility.

Particularly troublesome are the definitions of risk used in the various models. The Factual Information and Personal Gain/Loss models define risk as existing "out there" (albeit, in the case of the Personal Gain/Loss Model, "out there" as subjectively defined). Risk consists of probabilities of failure, magnitude of loss or some combination of these two aspects of the physical situation. The models thus rely on a stimulus-definition of risk (Vlek, 1987). The problem with such a definition is that it is practically intractable.

Communicator accommodation is better developed on the basis of a *response*-definition of risk, which is focused on internal psychological processes. The Values Model is compatible with such a response-definition. However, a major shortcoming of the Values Model is that it does not detail the specifics of the decision making process. This model also, obviously, does not provide much direction in those cases of communication in which differences in values are not a salient issue. The importance of a fully developed model of an audience's risk judgment and decision making, that begins with a response-definition of risk, is that it provides an understanding of the interactive processes between characteristics of the situation and the psychology of decision making.

4. A COMMUNICATION MODEL FOR RISK-ADAPTIVE DECISION MAKING

Our presentation of a decision-theoretical approach to hazard communication consists of two parts. First, we will deal with a transactional notion of perceived risk. Then we will make an attempt to combine classical (gambling) notions of decision making with more recent notions of decision making as dynamic-adaptive control of a demanding situation, in a 'Risk-Adaptive Decision Making' (RADeM) model of hazard communication.

4.1. Risk as insufficient controllability

Recent psychometric research on perceived risk of various technologies has taught us, among other things, that "riskiness" is strongly related to: (a) the image of a dreadful, catastrophic event, and (b) the (un)controllability of a risky activity. Psychologically, the concept of controllability seems to operate as a substitute for probability (e.g., Slovic, Fischhoff and Lichtenstein, 1980, 1984; Vlek and Stallen, 1981).

Perceived (lack of) control over an activity or situation someone is engaged in, serves as a key concept in modern theories of stress and coping with a demanding task. Especially in the work of Lazarus (e.g., Coyne and Lazarus, 1980) stress is defined in terms of the discrepancy between perceived task demands and assessed abilities to effectively cope with those demands.

Two emotional/behavioral response patterns are distinguished. First, when a person's assessed coping abilities do *not* meet the set of perceived task demands, the situation is experienced as threatening or risky: there is a potential for personal loss or harm. The corresponding dominant emotions are anxiety and fear. The ensuing behavioral pattern may involve passivity, defensive avoidance or (desperate) hyper-vigilance, and coping is generally ineffective.

In their study of perceived industrial safety in 'Rijnmond', a heavily industrialized area around the port of Rotterdam, Stallen and Tomas (1988) have demonstrated the importance of these patterns for understanding people's interest in hazard information. It was found that citizens with a secure or risk-accepting coping style held stronger beliefs that sudden accidents could be controlled than did those with highly vigilant or defensive patterns of coping. When an individual feels that (s)he is vulnerable to being victimized by a hazard and lacks a sense of control, various cognitive strategies, including defensive patterns such as those studied by Stallen and Tomas, might be used to adapt to the hazard (Janis and Mann, 1977).

An other emotional/behavioral pattern is "released" when assessed coping abilities are (just) higher than perceived task demands. The situation then is interpreted as challenging: there is a potential for mastery or gain. The dominant emotions are thrill and sensation. The behavioral activity reveals self-confidence and may be labelled as achieving, gaining, and instructive. Coping with the situation is generally effective. Changes in the situation and/or experiences with either effective or ineffective coping may lead to re-appraisal and further adaptive coping. Obviously, the provision of specific hazard information may contribute to this.

A risky situation, perceived as either threatening or challenging, is taken by this model to go through two stages of cognitive appraisal. In the *primary appraisal* stage it is inferred from the external task environment not necessarily in terms of explicit probabilities and magnitudes of possible losses, but in terms of a profile of (expected) task demands. During *secondary appraisal* the person assesses his or her available knowledge, abilities and resources to effectively deal with the set of expected task demands (or risk). Primary, environment-directed appraisal and secondary, abilities-directed appraisal generally parallels the distinction between stimulus and response definitions of risk. Neither of these two seems sufficient in itself as a basis for hazard communication. The combination of primary and secondary appraisal in a *transactional*, process-directed definition of risk (either threat or challenge) is needed to both understand and design hazard communication activities.

4.2. Assessing the acceptability of a risky (demanding) course of action

Some of the notions of risk-cost-benefit evaluation (cf. the Personal Gain/Loss and the Values models) can be usefully combined with the ideas about appraisal of coping abilities in potentially threatening situations. This has led us to sketch a structural model for judging the acceptability of a demanding course of action (e.g., living next to a complex technical installation).

Figure 2 presents a model of Risk-Adaptive Decision Making - or RADeM, elaborated from Vlek (1987, Figure 4), which pivots around four basic components: (a) desirable and (b) undesirable scenarios, (c) a profile of expected task demands, and (d) a corresponding profile of available means or abilities to cope with these demands.

The upper half of the model is inspired by classical decision theory with its emphasis on "external", uncontrollable chance factors, and observing a separation between positive and negative possible consequences. Together the desirable and undesirable scenarios lead to an (sub)aggregate judgment of the "average desirability of possible scenarios".

Figure 2: Communication model for Risk-Adaptive Decision Making (RADeM); decision framing (upper half) versus threat appraisal (lower half), and positive aspects (left half) versus negative aspects (right half). The periphery indicates highly specific considerations underlying analytical judgment; the center contains the overall (synthetic) judgements to be made.

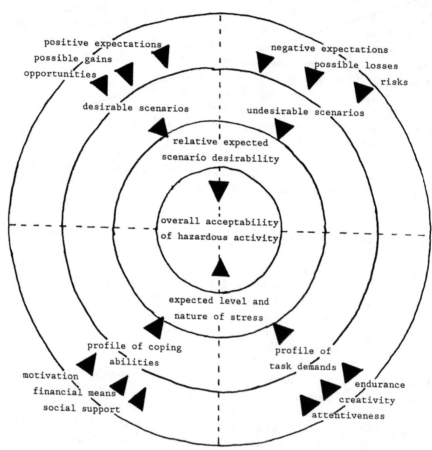

The lower half of Figure 2 is inspired by theoretical notions about stress from, and coping with expected task demands. It decomposes 'expected stress' (an other subaggregate judgment) into the expected profiles of task demands and coping abilities, respectively. The two subaggregate judgments: 'average scenario desirability' and 'expected stress', are then combined into an overall judgment of the acceptability of the course of action under consideration.

Following the Risk-Adaptative Decision Making (RADeM) model one would recommend hazard communicators to design their message - or rather their discourse with the audience - on the basis of both the negative and the positive aspects of the activity (the right and left halves of Figure 2, respectively), as well as on the basis of "external" - environmentally determined - and "internal" - personally determined - aspects (the upper and lower halves of Figure 2). The model implies that several things should be weighted against one another: desirable versus undesirable scenarios, coping abilities versus task demands, and - in the end - average scenario desirability against expected stress.

The model also reflects the possible variation between analytic and synthetic (intuitive) judgment, indicated with the help of the set of concentric circles. Highly analytic hazard communication goes all the way into the periphery of Figure 2 to spot and evaluate the various underlying aspects of the four major components of the model. The analytics' problem is to get back to the center for making the required overall acceptability judgment. This mode of acceptability judgment would seem to require the 'deep cognitive processing' discussed in Section 3.1 (see Assumption 2).

Highly synthetic hazard communication, and discourse, however, attempts to stay around the center of Figure 2. There, an overall acceptability judgment is 'nearby', but one risks overlooking or underestimating a number of detailed aspects that should underly a proper evaluation. This way of judging a risky or demanding activity's acceptability seems to go along well with a 'shallow' type of cognitive processing. Finally, Figure 2 is a 'satisficing', not a 'maximizing' model of decision making. It is applicable to separate courses of action and does not entail (nor exclude) a search for a larger set of feasible courses of action, and consequent comparative choice.

The RADeM model accommodates various elements from our earlier discussion. First, it implies that providing 'factual' (i.e., scientific/technical) information is necessary but insufficient. Also, factual information should pertain to positive as well as negative aspects of the activity under consideration. And the communicator may have to establish beforehand how much factual information is likely to be actually processed by the audience.

Second, the RADeM model suggests that risks, costs and benefits may be treated from an 'economic' perspective which largely follows from the policy maker's point of view. But they surely should also be treated from the audience's point of view and in terms of expected 'task demands' and available coping abilities and resources. That is, one should be aware that no economic benefit, however large, may compensate for excessive levels of task demands that would materialize if a hazardous technical facility is decided upon. The discussion on risks, costs and benefits - either from an economic or from a psychological point of view - should help the involved parties to become more explicit about basic values and principles that can guide their specific value judgments or acceptability criteria.

Third, the RADeM model seems to accommodate the controversy about the meaning of 'risk'. The statistical (and engineering) conception of risk as some function of the probability and the size of a possible loss or accident seems to fit best into the upper ('gambling') part of the model. In contrast, the psychological notion of risk as insufficient controllability fits into the lower ('control') part of Figure 2. Thus we recommend that hazard communicators realize that there may be different kinds of risk, depending upon the nature of the available underlying information base, which may be either 'external' or 'internal', and 'frequentistic' or basically 'nonfrequentistic' (Vlek, 1987).

Fourth and last, the RADeM model suggests that the relevant set of judgments and their underlying aspects may well be quite different for different parties involved in an acceptable-risk discussion. Here we suggest the adoption of the "marketing orientation" by Earle and Cvetkovich (1985, see also Section 3.1 above): designers of hazard communication programs should attempt to fit their approach and their specific messages to the various audiences that may be meaningfully distinguished. It is useless to present a packet of detailed information when the audience is incapable or unwilling to process that information. Also, it seems unbalanced to design that message on the basis of desirable and undesirable policy scenarios (upper half of Figure 2) when the audience actually is more concerned about issues revolving around expected stress from the activity under consideration (lower half of Figure 2). The interpretation of risk as the probability of an accident as against a lack of future controllability (perhaps reflecting a lack of faith in responsible experts) is of crucial importance here.

5. IMPLICATIONS AND CHALLENGES FOR HAZARD COMMUNICATORS

We have suggested that better accommodation in hazard communications about large scale technologies could be made if the model held of the audience was based on a clear understanding of how risky decisions are made and if a response definition of risk was assumed. The decision making model proposed here is based on these points. Central to this model is an understanding of the psychological processes by which an appraisal is made of threat and one's ability to cope with it. A situation is judged to be threatening if the individual concludes that (s)he is in trouble and (s)he or society does not have the means (or the intention) of coping with and controlling the trouble.

In sum, the RADeM Model, starting from a response-definition of risk, places central importance on judgments of controllability. The RADeM Model forces the communicator to recognize that risk communication (training and informing the public about the implicit riskiness of technological activities using technical and scientific information) is only one part of communications about risk which also includes credible non-technical information ensuring the public about the trustworthiness of the societal control of activities.

It follows from the model that better understanding of the audience (and thus more appropriate communicative accommodation) will be achieved by developing an understanding of the information processing underlying judgments of personal and societal control. Thus, the RADeM Model leads to an agenda of research which, due to the directions indicated by other models, has been largely overlooked.

On a practical level, the model indicates that communicators should recognize the central importance of threat and stress reactions and in some circumstances provide information that could aid in structuring the decision for the audience, identifying the sources of stress, and the means for controlling it. The greatest challenge to the hazard communicator of the RADeM Model is that the procedure used for making a technological decision will greatly affect judgments of the controllability of that technology. Thus the RADeM model allows for the conclusion that effective communication is composed of more than what is said in the message to the audience. To be effective, communications must occur in a context set by the use of what the audience deems to be appropriate decision making procedures. Suggestions on how this might be accomplished are offered in the concluding chapter of this volume.

REFERENCES

Baird, B. (1986). Tolerance for environmental health risks: The influence of knowledge, benefits, voluntariness and environmental attitudes. *Risk Analysis, 6,* 425-436.

Baird, B., Earle, T.C. and Cvetkovich, G. (1986). Public judgment of an environmental health hazard: Two studies of the ASARCO smelter. In: L. Lave (Ed.), *Risk assessment and management.* New York: Plenum.

Baram, M. (1984). The right to know and the duty to disclose hazard information. *American Journal of Public Health, 74,* 385-390.

Bezembinder, Th. (this volume). Social choice theory and practice.

Brouwer, H.C.G.M. (this volume). Current radioactive waste management policy in the Netherlands.

Chen, K. and Mathes, J.C. (this volume). Value oriented social decision analysis: a communication tool for public decision making on technological projects.

Commission of the European Communities (1982). Council directive of 24 June 1982 on the major-accident hazards of certain industrial activities. *Official Journal of the European Communities, No. L 230/1,* 5.8.82, Appendix I.

Covello, V., Von Winterfeldt, D. and Slovic, P. (1986). Communicating scientific information about health and environmental risks: Problems and opportunities from a social and behavioral perspective. In: V. Covello, A. Moghissi and V.R.R. Uppuluri (Eds.), *Uncertainties on risk assessment and risk management.* New York: Plenum.

Coyne, J.S. and Lazarus, R.S. (1980). Cognitive style, stress perception and coping. In: I.L. Kutash, L.B. Schlesinger and associates (Eds.), *Handbook on stress and anxiety* (pp. 144-158). San Francisco: Jossey-Bass.

Cvetkovich, G. (1978). Cognitive accommodation, language and social responsibility. *Social Psychology, 41(2),* 149-155.

Cvetkovich, G., Earle, T.C., Schinke, S.P., Gilchrist, L.D. and Trimble, J.E. (1986). Child and adolescent drug use: An information and judgment perspective. *Journal of Drug Education,* *17(4)*, 295-313.

Cvetkovich, G. and Keren, G. (1988). *Mental models and communicating* *environmental hazard information: Prospects, practice and* *problems.* Western Institute for Social and Organizational Research; Department of Psychology; Western Washington University; Report DM/RC 88-11.

Douglas, M. and Wildavsky, A. (1982). *Risk and culture.* Berkeley: University of California Press.

Earle, T.C. and Cvetkovich, G. (1985). *Risk communication: A* *marketing approach.* Report DM/RC 85-03. Western Institute for Social and Organizational Research; Department of Psychology; Western Washington University.

Earle, T.C. and Cvetkovich, G. (1986). Failure and success in public risk communication. In: The Air Pollution Control Association (Eds.), *Avoiding and managing environmental damage from major* *industrial accidents.* Pittsburgh, PA: Air Pollution Control Association.

Fishbein, M. and Ajzen, I. (1981). Acceptance, yielding and impact: Cognitive processes in persuasion. In: R.E. Petty, T.M. Ostrom and T.C. Brock (Eds.), *Cognitive response in persuasion.* Hillsdale, N.J.: Erlbaum.

Hisschemöller, M. and Midden, C.J.H. (this volume). Technological risk, policy theories and public perception in connection with the siting of hazardous facilities.

Janis, I.L. and Mann, L. (1977). *Decision making: A psychological* *analysis of conflict, choice and commitment.* New York: The Free Press.

Johnson, R. and Petcovic, W.L. (1986). *The risk of communicating risk.* Paper presented at the Annual Meeting of the Society of Risk Analysis, Boston, Mass. 9-12 November.

Kasperson, R.E. (1986). Six propositions on public participation and their relevance for risk communication. *Risk Analysis, 6(3),* 275-282.

O'Hare, M. (1977). Not on my block you don't. *Public Policy, 25,* 407-458.

Peltu, M. (1985). The role of communications media. In: H. Otway and M. Peltu (Eds.), *Regulating industrial risks; science, hazards and* *public protection.* London: Butterworths.

Petty, R.E. and Cacioppo, J.T. (1981). *Attitudes and persuasion:* *Classical and contemporary approaches.* Debuque, Iowa: W.C. Brown.

Phillips, L.D. (this volume). Requisite decision modelling for technological projects.

Rayner, S. (1986). *Risk and relativism in science for policy*. Oak Ridge Tennessee: Oak Ridge National Laboratory.

Rayner, S. and Cantor, R. (1987). How fair is safe enough? The cultural approach to societal technology choice. *Risk Analysis, 7*, 3-10.

Ruckelshaus, W. (1983). Science, risk and the public policy. *Science, 221*, 1026-1028.

Slovic, P. (1986). Informing and educating the public about risk. *Risk Analysis, 6(4)*, 403-416.

Slovic, P., Fischhoff, B. and Lichtenstein, S. (1980). Facts and fears: understanding perceived risk. In: R.C. Schwing and E. Albers Jr. (Eds.), *Societal risk assessment: how safe is safe enough?* (pp. 181-213). New York: Plenum.

Slovic, P., Fischhoff, B. and Lichtenstein, S. (1984). Decision theory perspectives on risk and safety. In: K. Borcherding, B. Brehmer, Ch. Vlek and W.A. Wagenaar (Eds.), *Research perspectives on decision making under uncertainty* (pp. 183-203). Amsterdam: North-Holland.

Stallen, P.J. and Tomas, A. (1988). Public concern about industrial hazards. *Risk Analysis, 8*, 237-245.

Thaler, R. (1985). *Perceptions of fairness in economic transactions*. Paper presented at the 10th Research Conference on Subjectibe Probability, Utility and Decision Making. Helsinki, August.

Törnblom, K. and Johnson, D.R. (1985). Subrules of the equity and contribution principles: Their perceived fairness in distribution and retribution. *Social Psychological Quarterly, 48(3)*, 249-261.

Tyler, T.R. (1986). Justice and leadership endorsement. In: R.R. Lao and D.O. Sears (Eds.), *Political cognition*. Hillsdale, N.J.: Erlbaum.

Tyler, T.R., Rasinski, K.A. and Griffin, E. (1986). Alternative images of the citizen. *American Psychologist, 41(9)*, 970-978.

Van der Pligt, J. (this volume). Nuclear waste: public perception and siting policy.

Van der Pligt, J. and Eiser, J.R. (1984). Dimensional salience, judgment, and attitudes. In: J.R. Eiser (Ed.), *Attitude Judgment* (pp. 161-177). New York: Springer.

Vlek, C.A.J. (1986). Rise, decline and aftermath of the Dutch "Societal Discussion on (Nuclear) Energy Policy" (1981-1983). In: H.A. Becker and A.L. Porter (Eds.), *Impact assessment today* (pp. 141-188). Utrecht (Neth.): Van Arkel.

Vlek, C.A.J. (1987). Risk analysis, risk perception and decision making about courses or action involving genetic risk: An overview of concepts and methods. In: G. Evers-Kiebooms, J.J. Cassiman, H. van den Berghe and G. d'Ydewalle (Eds.), *Genetic risk, risk perception and decision making* (pp. 171-207). New York: March of Dimes Birth Defects Original Article Series (A.R. Liss).

Vlek, C.A.J. and Stallen, P.J. (1981). Judging risks and benefits in the small and the large. *Organizational Behavior and Human Performance, 28,* 235-271.

Wilke, H. (this volume). Promoting personal decisions supporting the achievement of risky public goods.

DECISION MAKING IN TECHNOLOGICAL EMERGENCIES

URIEL ROSENTHAL AND MENNO J. VAN DUIN

Department of Public Administration
University of Leiden

1. INTRODUCTION

There is a peculiar bifurcation in popular notions of government. On the one hand, people unite in the "struggle against bureaucracy". Indeed, along with the universal fear of nuclear war, this is one of the few concerns shared by the people and political leaders in different political and social regimes. On the other hand, there is a rock-bottom belief that it is the state and its agencies which should protect the people from being struck by disasters, turmoil and terrorism. Moreover, if the state is unable to shield its citizens against such adversity, it should at least be the initiator of emergency management through quick and resolute decision making. It is the thesis of the present chapter that normal suspicions about governments' inadequacies might be profitably applied to disasters as well as to other political emergency situations.

At first sight, technological emergencies would seem to be excellent subjects for the application of rational and synoptic decision making methods. On closer inspection, however, nothing could be further from the truth. As policy analysts stress time and again, rational-synoptic decision making does not stand the empirical test in society-and-politics-as-usual (e.g., Braybrooke and Lindblom, 1963). There is an abundant amount of evidence showing that emergency management in particular does not meet the standards set by rational decision models (Comfort, 1988; Quarantelli, 1988; Rosenthal, 1988). In situations marked by a combination of threat and shortage of decision time, the best that can be hoped for is some kind of satisficing: reaching decisions that are 'good enough'.

Compared with other categories of emergency situations, technological emergencies score very high on the defining characteristics of extraordinary circumstances. If normal political circumstances already do not yield rational-synoptic decisions, and if emergencies set the

Ch. Vlek and G. Cvetkovich (eds.), Social Decision Methodology for Technological Projects, 277–295.

conditions for specific patterns of crisis decision making, then technological emergencies *a fortiori* must involve decision making processes far removed from pure-rationality and related models.

This chapter focuses on decision making at the time technological emergencies are happening. Usually, policy studies on technological emergencies attend to preventive measures, planning procedures, and policies in the aftermath of catastrophe. Here the main emphasis will be put on the role of the authorities during the unfolding of an actual technological emergency.

The kinds of emergency we have in mind are exemplified by such ill-reputed symbols of large-scale (near-)disaster like Seveso (1976), Three Mile Island (1979), Bhopal (1984), Chernobyl (1986) and Sandoz-Basel (1987). It should be clear that since the focus of the chapter will be on the responsibilities, decisions and nondecisions on the part of the authorities, less attention will be paid to the technical and operational levels of emergency decision making.

After considering in Section 2 some specific characteristics which define technological emergencies as crises, we present in Section 3 an analytical scheme for describing government decision making during crises. Section 4 examines to what extent six propositions pertaining to crisis decision making are applicable to actual technological emergencies. To conclude, Section 5 offers a summary and a number of recommendations for improved crisis decision making.

2. TECHNOLOGICAL EMERGENCIES AS CRISES

Whether natural or man-made, emergencies can, by their very nature, be considered as critical events in the history of a social system. They put everyday social and political reality into a "pressure cooker". Democratic structures and processes are challenged by the exigencies of acute or protracted threats.

Technological emergencies are social crises: serious threats to the basic structures or fundamental values of a social system which - under conditions of time pressure and uncertainty - require critical decisions (Hermann, 1972; Rosenthal, 1986, 1989). Up to now, crisis decision making as a label has been primarily reserved for international conflicts. But, following Dynes's (1974) suggestion, one may apply the notion of crisis decision making to a variety of extraordinary circumstances, including riots, civil strife and natural and man-made disasters. As a specific category of potential and actual calamities, technological emergencies may also be subsumed under this perspective.

In a way, technological emergencies must be high on the top list of *contradictiones in terminis*. Technology would seem to be the example 'par excellence' of man's control over nature. It would seem to be at odds with all those negative connotations of emergencies summarized as the "un-ness" of the problem: unmanageable, unexpected, unprecedented, uncertain, unaware, unready, unscheduled (Hewitt, 1983).

It is no wonder that there is a wide gap between the paradigm of technological control and the perspective of crisis decision making. In contradistinction to the notion of technological progress, the crisis concept emphasizes the idea of failure - the idea of things running out of control. However, there is no reason to foreclose the possibility that technological emergencies will occur. Indeed, acute technological emergencies contain many ingredients of the just mentioned "un-ness" of the problem.

In stressing the contradiction between control and crisis one should be aware that crisis decision making goes beyond those stages in the process that yield to a significant degree of control. At the moment of crisis, preventive policies and emergency planning do not suffice anymore. Control as prevention is irrelevant. In its quality of a complete elimination of threat and misery, control would furthermore amount to a sudden capacity to fully cope with what, up until that moment, must be characterized in terms of the above mentioned "un-ness" of the situation.

In case of an acute emergency, decision makers do not have the time and often lack the tools to prevent substantial damage from taking place. They may be lucky and strong enough to set priorities and do their utmost in containing the damage. It may be more important to prevent a medium-scale technological disaster from assuming dramatic large-scale proportions than to strive for the impossible. When a severe threat has become a reality, it can not be undone. Risky decisions have to be made without the kind of thorough risk assessment found in books (Griffiths, 1981).

Technological decisions have increasingly become the target of intense controversy: "For those who are trying to establish a new principle ... each occasion for decision takes on aspects of a crusade" (Thomson, 1976, p. 37). Technology is not just a product of scientific efforts. It is as much an outcome of conflict and competition between various interests and institutional spheres. Indeed, scientific institutions are said to be "poorly prepared" to stand firm in the political and institutional fights on the regulation and deregulation of technologies (Krimsky, 1984).

Thus technological emergencies - uncontrollable technology - are highly political events, containing a tremendous conflict potential. A crisis decision making perspective should predict the prevalence of distinctly political processes during periods of acute threat or uncertainty. Crises are political events. In crisis situations, one may observe specific patterns of decision making which invariably are highly political in format and content. Before discussing these, let us consider three aspects of the political situation in which a technological emergency may occur.

Private versus public sectors. It is practically inevitable that emergencies occur in a climate of existing tension between the private and the public sectors. Corporations, especially those dealing with complex high-technological processes, keep close contact with executive government agencies concerning their 'external safety'. It is not unlikely that certain industrial facilities do not work according to official regulations. Contact with government costs time and there is a chance that its agencies require modifications of the production processes. And entrepreneurs may wonder why *their* company, and not other similar companies, is closely inspected.

In its turn, public administration shows ambivalence regarding the industry. On the one hand the safety of people living in the area and of the workers must be guaranteed. On the other hand, the representatives from the public sector know all too well how crucial these 'hazardous' companies may be for the economy. A lot of industrial activity in the area is attractive because of profit taxes, income and work.

Multi-agency process. During a technological emergency government is omni-present. Local, provincial and national governments may all be active in the coordination of assisting, warning and informing the people involved. At the operational level, company services such as fire brigades and assistance from various government agencies (fire departments and police and ambulance services) may have to collaborate. Local agencies often call in assistance from neighbouring towns and specialized provincial and national teams. All in all, there will be many people and organizations involved. Not all of the activities will run parallel to one another.

Post-Crisis Politics. Research is frequently done after a crisis. The people and organizations involved are assessed with regard to how well they functioned during the crisis. More often than not it turns out that certain procedures were not carried out, supervision fell apart or important safety measures were ignored. Many problems and errors were previously known. After the catastrophe with the Challenger space-shuttle, for example, it was disclosed that some operators had warned time and again against a take-off at such low temperatures. The government finds itself in a remarkable double-position since it is often the first to be accused of failures regarding the prevention of accidents.

3. GOVERNMENTAL CRISIS DECISION MAKING: AN ANALYTICAL FRAMEWORK

Psychologists are familiar with the intricate distinction between objective and subjective risk. A similar distinction may apply to the notion of crisis. Each of the defining features of crisis (serious threat, urgency, or uncertainty) lends itself to diverging perceptions. What is a severe threat in the eyes of some actors, may be an asset or indeed a splendid opportunity to others. The sens of urgency among some authorities may be at odds with the laconism of others. Uncertainty may reach beyond standardized measures of informational shortage; it may also involve subjective estimates of the solidity of incoming information.

In order to avoid confusion between an objective and a subjective notion of crisis decision making it is necessary to be quite specific about the conditions that determine the dimensions of threat involved in the motions and actions of a threat agent, as well as about the autorities' awareness that crisis decision making is pertinent to the situation. To clarify the process of crisis decision making we will consider five phases or conditions that seem necessary for effective decisions to be made. It is only in passing from the fourth to the fifth phase that the objective conditions for crisis decision making will have prepared the ground for actual activities to that effect on the part of the authorities.

3.1 The social system (its basic structures, fundamental values, or crucial norms) should be exposed to a **serious threat**.

The central point of the first phase is the threat to the *existing* social system. This threat may affect widely divergent system components. Whether private or public, these different components will often be threatened either at the same time or consecutively.

The rapid socio-economic developments over the last decades have been mostly due to industrial and technological growth. Society generally has faith in new technology but it has also become dependent upon it. Catastrophes such as those in Bhopal (1984) and Chernobyl (1986) raise questions as to social, economic and technological priorities. It is quite conceivable that our social system just can not handle certain high-risk technologies. Technological emergencies cause damage to public trust in technicians and in scientific developments. From this stems an additional threat to values: the existence of self-destroying technologies.

3.2 It is necessary for the **social system** to **respond** to the threat.

The second phase refers to the *continuence* of the social system. To ensure this, it is of primary concern to reduce or eliminate the immediate threat, or, if that would be impossible, at least to protect essential system elements. The need to divert the danger is unquestionable.

Because the foundations of a social system may be affected during a technological emergency, a reaction from the government is necessary. Abandoning the present level of technological and industrial development and thereby effectively abandoning the current standard of living we enjoy, is undesirable. However, technological emergencies may call for specific and radical changes within the system. Perrow (1984) suggests that systems that can not be salvaged, such as nuclear weapons and nuclear power, should be abandoned because their inevitable risks outweigh any reasonable benefits. Some other systems could be made less risky with considerable effort (e.g., some kinds of marine transport and DNA research) or with modest adjustments (e.g., chemical plants).

3.3 It is necessary for **government authorities** to take **decisions** in order to avert or contain the threat.

Here the point of reference is the capability of the elites, the political and bureaucratic authorities in particular, *to respond* to the challenge of a serious threat. Having accepted the postulate of the system's persistence, we now face the question whether capabilities and responsibilities are adequate to ensure this. As we have already seen, there is no guarantee that the activities of responsible authorities will necessarily result in effective crisis management.

3.4 It is necessary for government authorities to decide **promptly** in order to avert or contain the threat.

The fourth phase entails the necessity of *prompt* decision making by government authorities. Crisis decision making is decision making under pressure. A crisis implies a combination of accelerated decision making and critical decisions. Within a few hours, minutes or even a split of a second, decisions must be made that indeed involve matters of life and death - to be or not to be. Wiener and Kahn (1962) describe the decision making context as follows: decisions and other negotiations are immediately required; the decisions may have decisive implications for the social system; the pressure for time increases; these aspects interlock with each other.

Experience shows that it is the hurriedness in negotiations during a technological emergency that gives rise to many problems. In a number of substantial emergencies, notably those involving uncontrolled releases of toxic substances, explosions and nuclear reactions, every minute is of vital importance. Still, companies sometimes will wait as long as possible to reveal the threat of danger. Much precious time may pass while the problems are increasing. In many cases the extent of the danger is underestimated and the impression given that things are not that bad after all. For example, following the release of a toxic gas cloud in Seveso, Italy, the company involved initially revealed that a cloud of herbicide had spread over the area around the factory. Only days later did it become clear that it had been a cloud of dioxin (see Hay, 1982; Lagadec, 1982).

*3.5 Government authorities **actually engage** in crisis decision making.*

It is only in the fifth phase that we see authorities actually *engage* in crisis decision making. This phase involves the move from the analytical urge for promptly taken critical decisions, to the action-oriented awareness on the part of the authorities that these should, indeed, be taken. Authorities take a real threat to the social system seriously. They are determined that the threat will not be permitted to lead to a collapse of the system. They are realizing that *they* are responsible for taking critical decisions in order to avoid the danger.

In technological emergencies this same sequence is applicable, mutatis mutandis, to managers of firms as well as systems operators. Instead of embracing the idea of an automatic, mechanical reaction to technological emergencies, our analytical framework shows that many obstacles must be removed before governments and corporations "do what they have to do".

4. PROPOSITIONS ABOUT THE NATURE OF CRISIS DECISION MAKING

Let us now consider six characteristics that we propose are essential for understanding decision making during (technological) emergencies. These propositions are inferred from observations in the fields of international relations (with the cornerstone of the nuclear crisis), the sociology and politics of disaster, studies in organizational turbulence, and analyses of collective stress.

4.1 In crisis situations decision making becomes increasingly centralized.

Technological emergencies are no exception to the rule found to apply to international crises and to a variety of domestic crises. When, in spite of reassuring statements about the level of technological control, a large-scale technological disaster or near-disaster takes place, national authorities may step in to show their resoluteness vis-à-vis the state or local authorities who are closer to the site of the catastrophe and, consequently, the first to be blamed. Institutional arrangements may help to reinforce the tendency to centralized decision making. For instance, an emergency may be declared a "federal disaster".

We are talking politics here. Technological emergencies may offer an opportunity for relatively safe media exposure to the central authorities: "we give them all the help we can." The allocation of personnel and emergency goods by the central government suggests that it is the local authorities or private companies that have manoeuvred themselves into a catastrophic situation. The central authorities may present themselves as the "good guys" who are willing to assist the authorities at the site of the disaster in regaining control over the situation. The U.S. National Aeronautics and Space Administration, for example, was the "bad guy" after the space shuttle crash; the American President expressing the feeling of grief and sorrow in the name of the American people, represented the "good guy".

The Disaster Act of the Netherlands formally assigns primary responsibility for disaster management to the local authorities, while the role of the provincial and central authorities appears to focus upon giving assistance to local relief agencies. However, this is only part of the picture. The basic thought behind Dutch disaster legislation actually is that a real disaster requires assistance from outside the local domain. It is up to the Provincial Governor to intervene whenever (s)he has the impression that a disaster is not manageable by the local authorities. It is up to the Secretary of the Interior to take decisions in case of catastrophes involving more than one province. This may easily induce the provincial and central authorities to interfere with any disaster, irrespective of its impact and consequences beyond the local area. External assistance seems to imply influence from higher levels of government.

One could argue that this reflects a realistic approach to crisis management. But it probably also results from the ingrained belief that "victims can not cope without our help" - and that, as a corollary, communal authorities and agencies had better leave the vital decisions to an other, indeed higher, level of government (Cuny, 1983; Rosenthal, 1988).

4.2 In crisis situations bureaupolitics flourish.

According to one of the dominant models of bureaucratic politics (Allison, 1971), bureaucratic agencies (both public and semi-public bodies) try to increase their power and attempt to promote their organization's goals and identity. Success in promoting the organization's essence supposedly depends on its capacity to propagate and exhibit expertise and high morale, if possible under difficult circumstances (Halperin, 1974; Gray and Jenkins, 1985).

Consensus and solidarity are part of the official ideologies and planning schemes for coping with technological disasters (Cantelon and Williams, 1982). However, the reality of emergency decision making shows that it is naive to think that political authorities and public agencies suddenly lose interest in their position on the ranking orders of power and prestige. For those authorities and agencies who have a special responsibility for managing acute emergencies, success and failure may also involve their own *raison d'être*. Emergency and emergency-relevant organizations will make every effort to fight their way into the decision arena.

If consensus and solidarity have a role to play in this context, it relates primarily to a quite specific dimension of bureau-political rivalry and competition. Inter-agency conflicts, which sometimes amount to televised quarrels between agency leaders, may include professional differences of opinion. The bureau-politics of emergency decision making may be of such intensity because considerations of power and prestige get mixed up with the sincere belief each agency has in the soundness of its own decisions and actions.

Bureau-political conflicts not only occur during but especially after a crisis. There is no activity which, by its specific substance, will never be the subject of competition and conflict. Helping the victims of a disaster often involves a bitter contest between the Red Cross, the Salvation Army, and an intriguing variety of welfare institutions. It has been aptly called the war of the good Samaritans.

For example, in the wake of the nuclear reactor accident at Chernobyl, directly after the danger of radiation was over in the Netherlands, reproaches were thrown back and forth. A top official of one Ministry publicly suggested that the Ministry of the Interior - responsible for disaster management in Holland - was not prepared for, or in a position to react adequately in a disaster situation of a certain size.

We may conclude that crisis decision making with regard to technological emergencies is not just a matter of coordination among the many actors involved. With Quarantelli (1988) one should indeed recognize that inter-organizational coordination may become an additional problem rather than a distinct solution. In a similar vein it can be argued that smooth rules and operating procedures will only serve the decision makers to the extent that they have become part of the decisional culture. This observation boils down to a definite demand for communicative planning, training and drilling in the preparatory phase of emergency management (Scanlon, 1989).

4.3 In crisis situations formal rules and command lines are less strictly respected.

Situational leaders and technical advisers tend to advance into decision making positions. With regard to technological emergencies the role of technical advisors in decision units seems to be of particular interest. If the focus of the decision making process is at safe distance from the site of the disaster, the authorities may claim that events have taken place beyond their power domain. Despite the urgency of the situation - and sometimes due to time pressures - they may consult technical experts who, similarly, have no direct relation to what has happened "over there".

But from a political perspective, things may look quite different. They may, indeed, look like a silent conspiracy between failing authorities and equally deficient technicians. Both political and technical authority - authority being understood as legitimate power - can be ruined by a single (near-) disaster: "If experts are wrong, then this may not have been the one time in three hundred years, but the first one of many, many times over three hundred years" (Perrow, 1984). Eventually, authority turns out to be seen as unreliable (Friedrich, 1963).

It is here that situational leaders step in. In a situation of dramatic threat formal power tends to yield to 'genuine' authority, whether technically or politically or socially oriented. Genuine, trusted and accepted, authority may emanate from anywhere, transgressing the ordinary hierarchical obstacles and surpassing a variety of carefully planned emergency procedures. Eventually, such authority may clash with the coalition of formal power and *expertise manqué.*

Once again, we have a clear example of the political processes at the core of crisis decision making during a technological disaster. What at first sight looks like a splendid opportunity for spontaneous consensus and solidarity, ends as a political phenomenon of the first order - not only in the aftermath, but also at the very moment of potential and actual threat.

4.4 In crisis situations decision makers tend to rely on "trusted, liked sources" (Milburn, 1972).

Technological emergencies tend to come as a shock to decision makers and their advisers; one should not be over-optimistic about the mitigating effects of pre-emergency planning activities and simulation games. When a technological emergency occurs, reason and balanced assessment easily give way to a *'Freund-Feind'* contradiction and biased information processes. Decision makers tend to prefer information from allies and supporters. They tend to reason away information offered by critics. They pursue a hardball approach to counter-expertise, however important such counter-expertise might be for the termination of acute threat (Etheredge, 1985). They do not welcome "self-reliant" solutions on the part of citizens and private associations. There is a tendency toward hypervigilance and, as a logical follow-up, post-decisional regret (Janis and Mann, 1977), which, for example, may induce a preference for forced evacuations at a rather early stage of emergency (Perry and Mushkatel, 1984).

4.5 In crisis situations decision makers tend to fill the information gap by analogy to previous crises.

History is more or less divided into time periods with the help of crises and other drastic events. These unusual circumstances determine the line from the recent past to the future for both the participants and the observers. Crises are both a collective and an individual reference point. For most inhabitants of the Dutch province of Zeeland, for example, the flood disaster of 1953 has, as reference point in time, a greater significance than did World War II. By implication, an *ad hoc* crisis unit consisting of authorities with differing crisis experiences, may face equally differing mental orientations to the emergency at hand.

When a crisis occurs, people have the inclination to refer to 'the last time', a previous crisis of more or less the same type and degree. One tends to look back on the last war, the last disaster, the last huge conflict. Sometimes people will be reminded of a dramatic disaster of type *x* when type *y* occurs. In the Netherlands, the rapid succession of the Bhopal, Mexico City and Sao Paolo disasters fully determined the emergency management process. Subsequently, the Chernobyl calamity caused a dramatic change of priorities. Imagine that the Netherlands would have faced an emergency of a totally different sort in those respective periods. Then, it would have been quite difficult for the decision makers to reorient their attention to that unexpected event.

A revealing version of this process can be found with the Dutch planning process regarding nuclear incidents. Instigated by the Chernobyl shock, all planning schemes initially read as a detailed operation to prevent such a calamity from occurring on Dutch territory. The simple fact that in the Chernobyl case collective stress derived from a particular sort of disaster in an other country seemed to be of secondary importance. When, after some time, the world heard of the nuclear waste affair in Goiana (Brazil), the planning schemes suddenly embraced this kind of emergency as a most credible incident to come.

Imagine, then, that Chernobyl would have been followed by a large-scale emergency which would have looked like it, but at closer inspection would have been essentially different. Imagine that shortly after the nuclear waste affair in the Goiana hospital a superficially similar, but actually different case would have occurred. One may indeed argue that under those circumstances decision makers would easily tend to reason by false analogy and would, consequently, be inclined to engage in deficient decision making.

Individuals as well as organizations can learn from earlier disasters. This kind of learning links up remarkably well with the type of learning in the bureaucratic context. It amounts to learning by precedent rather than learning by doing. The tendency to bureaucratize the learning process through formal evaluation procedures and officially accorded revisions of emergency plans and scenarios, is hard to overcome. Learning by precedent definitely appears to be more pervasive than learning by doing.

Such learning is defective to the extent that it overemphasizes similarities between successive technological emergencies, thus neglecting the changes which have inevitably taken place since previous crises. It should be immediately noted that those changes may be of different kinds. Sometimes those responsible for emergency management draw correct lessons from (near-) disasters; sometimes they do not. Unfortunately, there are some reasons for a fairly pessimistic view in this respect. There is a strong inclination among authorities and governmental agencies to "bureaucratize" the lessons which can be drawn from previous emergencies. Emergency scenarios become increasingly sophisticated; procedures are rendered more 'perfect' than before. Creativity and improvisation submerged in the officials' search for certainty and risk avoidance.

4.6 In crisis situations decision makers are inclined to yield to groupthink (Janis, 1972).

Unanimous decisions resulting from collective deliberations may give the impression of resoluteness, but they may in fact emanate from a collective pattern of defensive avoidance (Janis and Mann, 1977). One prefers to place the responsibility for decision making on 'the group'. Conforming to the group's norms guards against confrontations. He or she who thinks that unanimous decisions in such situations are the best decisions, does not realize that unanimity may be an escape for the recognition of individual responsibility.

Three factors are conducive to groupthink: (a) the group dynamics in connection with collective decision making, whereby the individual participant has to yield control over his or her own judgement and thinking (see Bezembinder, this volume), (b) social control, whereby a deviating opinion becomes identified with obstruction, sabotage and 'out-group' conceptions, and (c) anticipation of criticism and of eventual fiascos whereby individual responsibilities give way to diffuse 'group responsibility'.

Again, technological emergencies seem to be "excellent" opportunities for decision makers facing such situations to become victims of groupthink. Acute technological emergencies testify to a sudden loss of control over a crucial element of modern life. The pressures to take the single decision that will restore the confidence in the governmental handling of technology will be enormous. Knowing that they are compelled to take a critical decision with regard to "processes that can be described, but not really understood" (Perrow, 1984), decision makers may tend to hide behind the relatively safe protective shield of the group.

5. LESSONS TO BE LEARNED

The language of politics is a treacherous one. Many of its basic terms reflect political wishes rather than political reality. Although they may lack legitimacy, officials are called authorities. Although they may lack the qualities for leadership, politicians are quickly called political leaders very easily. Although one may agree that it has gradually lost most of its traditional functions, the political system continues to be labeled as the state. Although bureaucratic politics has become a dominant trait of public administration, public interest and the common good remain popular in political discourse. Last but not least, although we all know that political decision making is a matter of muddling through, and of satisficing at best (Braybrooke and Lindblom, 1963), we often associate government with some kind of rational process. In contrast to the market, government must involve some degree of planning and design.

This kind of wishful thinking may assume dramatic proportions with regard to emergencies, including technological ones. No modern state may, by itself, be able to protect its citizens against external aggression; a super-power is only powerful by the grace of mutual deterrence. But may we not expect the state to protect us from technological disaster? And if we are to be disappointed in this respect, may we not expect the state and its officials to take appropriate action in times of emergency? These would not seem to be exaggerative questions - the more so because we can not expect citizens to stay cool-headed when technological disaster strikes.

The reality of governmental decision making during technological emergencies, however, is at odds with this perspective. In a number of ways, emergency decision making exhibits serious deficiencies. Authorities and governmental institutions, including emergency and emergency-relevant agencies, may fall short of even modest criteria of proper decision making. On the contrary, citizens and private organizations may do much better than official views would suggest (Quarantelli, 1978; Cuny, 1983).

6. RECOMMENDATIONS

1. The first thing to keep in mind is that crisis decision making is decision making under time pressure. This characteristic of decision making makes it difficult, maybe even impossible to

employ a few well-known decision techniques. Written preparations of decisions, as well as referring the decision to a committee, wastes too much valuable time. Consulting with a diversity of interest groups takes too long; reaching agreements with these groups takes even longer. To first ask the opinions of experts and then weigh and evaluate their advice for a specific decision can be a cumbersome process.

Thus, a crisis situation does not lend itself readily to rational methods of decision making. To investigate all possible alternatives, anticipate all possible consequences of each alternative, establish a preferred order of alternative actions, and then determine the best solution: even under normal circumstances this is more an ideal than a reality. Norms and reality confront each other and force decision makers to go for satisfactory decisions rather than optimal ones.

According to Simon's theory of 'satisficing' (1957), one sensibly accepts as a starting point the fact that human and organizational capacities are constrained by a bounded rationality. Many problems require a satisfactory solution: one simply sets a level of aspiration and chooses the first alternative that reaches this level. One must resign oneself to the awareness that it may not have been the ultimately best solution.

2. Time pressure during a crisis is unavoidable. Because there is so much to anticipate, sometimes irreversible decisions must be taken. Yet it is sometimes better to take no decision than a last-second decision. Buying time may be a wise strategy - even if the situation would seem to demand immediate action.

3. However limited the time available for emergency decision making may be, an attempt should be made to discuss the undertaken procedures collectively. Not only must one pay attention to the consequences of procedures, but also to their actual implementation. Just as in everyday situations, there is a discrepancy between intended and final results of policy. Since decision making often ends up decentralized, advice from experts is desirable. Including experts in the decision can help in reducing this discrepancy.

4. Bureau-political conflicts are also unavoidable. Realizing this, one should not expect these conflicts, during a crisis, to disappear like

snow in the sun. All too often it is assumed that organizations, in times of need, will operate harmoniously and in a brotherly manner. Regular consultations between organizations as well as exercises in disaster management could contribute to a better understanding of each other's responsibilities and to improved insight into the problems and limitations of the various organizations.

5. During crises, it is imperative that information and knowledge be gathered from as wide a variety of sources as possible. Special attention must be paid to other than trusted, liked sources. By involving outsiders and experts of countervailing organizations in the decision, allowing different groups to operate independently of one another, and by considering possible alternative scenarios, groupthink may be avoided.

6. During a crisis, the decisions taken and procedures used must be recorded as accurately as possible. This facilitates later evaluations. Administrators and industrial leaders must be held ultimately accountable for the policy chosen. This would prevent the possibility of hiding behind field operators or operational services (such as the fire departments). It is all too possible to place the blame for technological emergencies on the mistakes of a few people.

7. Evaluations are not only important for identifying actual responsibilities. They must first and foremost be seen as learning instruments. Those directly involved as well as others may strongly profit from such evaluations. In fact, a very important latent function of post-crisis evaluation is to hold the attention of a great many authorities, who may feel inclined to get back to business as usual as soon as possible. But one should recognize that it will never be possible to really raise the level of concern and interest among politicians and bureaucrats as to the unpleasant, nasty kinds of events called technological emergencies and crises. In that sense, politicians and bureaucrats are very human after all.

REFERENCES

Allison, G.T. (1971). *Essence of decision*. Boston: Little, Brown and Company.

Bezembinder, Th. (this volume). Social choice theory and practice.

Braybrooke, D. and Lindblom, C.E. (1963). *A strategy of decision;*

policy evaluation as a social process. New York/London: The Free Press, Colllier/MacMillan (1970 ed.).

Cantelon, P.L. and Williams, R.C. (1982). *Crisis contained*. Carbondale (Ill.): Southern Illinois University Press.

Comfort, L.K. (1988). Designing policy for action: the emergency management system. In: L.K. Comfort (Ed.), *Managing disaster: strategies and policy perspectives* (pp. 3-21). Durham (N.C.): Duke University Press.

Cuny, F.C. (1983). *Disasters and development*. Oxford: Oxford University Press.

Dynes, R.R. (1974). *Organized behavior in disaster*. Columbus (Ohio): Heath-Lexington.

Etheredge, L.S. (1985). *Can governments learn?* New York: Pergamon Press.

Friedrich, C.J. (1963). *Man and his government*. New York: McGraw-Hill.

Gray, A. and Jenkins, W.I. (1985). *Administrative politics in British government*. Brighton: Wheatsheaf.

Griffiths, R.F. (Ed.).(1981). *Dealing with risk*. Manchester (U.K.): Manchester University Press.

Halperin, M.H. (1974). *Bureaucratic politics and foreign policy*. Washington D.C.: Brookings.

Hay, A. (1982). *The chemical scythe: Lessons of 2, 4, 5, -T and dioxin*. New York: Plenum Press.

Hermann, C.F. (Ed.).(1972). *International crises*. New York: The Free Press.

Hewitt, K. (Ed.).(1983). *Interpretations of calamity*. London: Allen and Unwin.

Janis, I.L. (1972). *Victims of groupthink*. Boston: Houghton-Mifflin.

Janis, I.L. and Mann, L. (1977). *Decision making; a psychological analysis of conflict, choice and commitment*. New York: The Free Press.

Krimsky, S. (1984). Regulating Recombinant DNA Research. In: D. Nelkin (Ed.), *Controversy: politics of technical decisions*. London: Sage.

Lagadec, P. (1982). *Major technological risks: an assessment of industrial disasters*. Oxford: Pergamon Press.

Milburn, T.W. (1972). The management of crisis. In: C.F. Hermann (Ed.), *International crises*. New York: The Free Press.

Perrow, C. (1984). *Normal accidents; living with high-risk technologies*. New York: Basic Books.

Perry, R.W. and Mushkatel, A.H. (1984). *Disaster management: warning, response and community relocation.* Westpoint (Conn.): Quorum Books.

Quarantelli, E. (Ed.).(1978). *Disasters.* London.

Quarantelli, E.L. (1988). Disaster crisis management. *Journal of Management Studies, 25,* 373-384.

Rosenthal, U. (1986). Crisis decision-making in the Netherlands. *The Netherlands' Journal of Sociology, 22,* 103-129.

Rosenthal, U. (1988). Disaster management in the Netherlands: planning for real events. In: L.K. Comfort (Ed.), *Managing disaster: strategies and policy perspectives.* Durham (N.C.): Duke University Press.

Rosenthal, U., Charles, M.T. and 't Hart, P. (Eds.).(1989). *Coping with crises: the management of disasters, riots and terrorism.* Springfield (Ill.): C.C. Thomas.

Scanlon, J.T. (1989). Planning and crisis decision making: the Mississauga case. In: U. Rosenthal, M.T. Charles and P. 't Hart (Eds.), *Coping with crises: the management of disasters, riots and terrorism.* Springfield (Ill.): C.C. Thomas.

Simon, H.A. (1957). *Models of man.* New York: Wiley.

Thomson, I.T. (1976). The Tocks Island dam controversy. In: L.H. Tribe, C.S. Schelling, and J. Voss (Eds.), *When values conflict.* Cambridge (Mass.): Ballinger.

Wiener, A.J. and Kahn, H. (1962). *Crisis and control.* Harmon on Hudson: Hudson Institute.

SOCIAL DECISION MAKING ON TECHNOLOGICAL PROJECTS: REVIEW OF KEY ISSUES AND A RECOMMENDED PROCEDURE

CHARLES VLEK AND GEORGE CVETKOVICH

Department of Psychology
University of Groningen

Department of Psychology
Western Washington University, Bellingham

So far in this volume the reader has been led along crucial theoretical themes, useful methodological approaches, and a number of practical findings and experiences. In this concluding chapter we will cut across the above material by addressing a list of topics that are essential in developing a social decision methodology applicable to technological projects. Many of the topics were highlighted during discussions among the initial workshop participants in 1986. Others were instigated by later contributors or came up when we considered the set of chapters as a whole.

Our list begins with a recapitulation of the nature of socio-technological decision problems, a critique of technical risk analysis and risk management, and some notes about the role of experts. We then go into several issues concerning individual and social decision making. The possibilities and limitations of computer-supported decision making are considered in view of criteria for 'better decisions'. After having argued the need for increased public involvement, we stress the close relation between 'risk communication' models and social decision approaches. The chapter ends with a proposed outline of a practical social decision procedure.

1. CHARACTERISTICS OF LARGE-SCALE TECHNOLOGY (LST) DECISION PROBLEMS

Classical decision theory is based upon well-defined choice problems in which the alternative options and their relevant consequences are clearly given beforehand. Thus comparative evaluation and overall

Ch. Vlek and G. Cvetkovich (eds.), Social Decision Methodology for Technological Projects, 297–322.
© *1989 by Kluwer Academic Publishers.*

preference judgment are the major tasks facing a rational decision maker. The quality of decision making is determined by methodical coherence and (for repetitive similar problems) by the value of the average outcome. This hardly applies to decision problems related to large-scale technology (LST) whose nature seems to defy rational decision making according to any normative model.

LST decision problems are only partly *evaluation and choice* problems. They are also design problems, and as such they are primarily *structuring* problems. Often a proposed design of a technical facility is being presented for evaluation. A negative judgment on the design usually leads to a subsequent improved design which is then put up for an other round of evaluation. A 'go' decision is only made at the end of a series of design trials when the latest version meets the acceptance criteria set by the responsible authorities. Thus we are dealing here with a dynamic rather than a static decision process, regulated and eventually terminated by a 'satisficing', not a 'maximizing' decision rule.

LST decision problems, then, are dynamic. They are also hard to define and this is largely due to their relatively novel, long-term, wide-spread and difficult-to-control possible consequences and effects. Therefore, 'experts' are actually only partial experts, and thus problems of public distrust may arise. The authority to decide upon, and to control the implementation of a technological option lies in the hands of few people, whereas possible consequences befall many. This may create problems of unfairness, both with respect to participation in the decision procedure and in the distribution of benefits, costs and risks. LST also brings about a motivational imbalance in the evaluation of (short-term) benefits against (long-term) risks.

Taken together, these characteristics make it difficult to communicate effectively among different involved institutions, parties and groups. Communication problems are only enhanced by the fact that those involved often differ in knowledgeability, and in the basic principles and goals which underly specific value judgments. See also our 'Introduction and Overview' chapter.

A specific set of problems besetting the technological society has to do with the required long lead time for implementing new large-scale technical facilities. Collingridge (1980, p. 19) attributes to this a 'dilemma of control' which he summarizes thus:

"... attempting to control a technology is difficult and not rarely impossible, because during its early stages, when it can be controlled, not enough can be know about its harmful consequences to warrant controlling its development; but by the time these consequences are apparent, control has become costly and slow."

To properly deal with the threat of irreversibility (the all-or-none character of large-scale technology) we need, on the one hand, careful and balanced technology assessment, (re-)design and decision making procedures. On the other hand, knowing that human designers, decision makers and operators will always have limited attention, skill and imagination, one wonders to what extent a general warning sign: 'please keep it flexible' should be affixed to any technological enterprise that could end in something having the large-scale, all-or-none and irreversible nature that Collingridge (1980) and also Sjöberg (1980) and Perrow (1984) have been warning against.

Given the dynamic, ill-defined, motivationally imbalanced, irreversible, and multi-party nature of LST decision problems, there *is no* rational decision theory that could provide a methodological recipe to solve such problems in a guaranteed optimal way. To the extent that the decision situation is ill-defined (due to lack of information about alternatives and/or consequences, or to the impossibility of valid probability and/or value judgments), individual rationality could not be defined. And in so far as various involved parties each hold on to their own specific preference order, an acceptable social decision could only be approached by compromising on collective rationality, equal participation (or power), and decisiveness (see Bezembinder, this volume).

2. RISK ANALYSIS AND RISK MANAGEMENT

In the 'Introduction and Overview' chapter we noted that the 'catastrophe potential' of large-scale technology has raised critical questions about the meaning of risk and of 'acceptable risk'. In the old days quite a few technical experts cherished the convenient view that risk was something 'out there' and measurable as a single-dimensional variable. They thus used a propensity definition of risk which was limited to the probability of an event having unwanted consequences.

A major weakness of classical risk analysis resides in the practical impossibility to meet the essential conditions required for the frequentistic probability concept to apply: (1) that there is a well-defined sample space of possible accidents and accident opportunities; (2) that there is a sufficient number of repeated observed events (accidents); and (3) that past and future observation conditions will be substantially similar.

Since psychologists and decision theorists joined in on the methodological debate (some publics had already protested on intuitive grounds), it became clear that one could utilize various formal definitions of risk. Moreover, psychometric research on comparative risk judgment was revealing the operation of various cognitive dimensions of perceived riskiness. Table 1 lists the main formal definitions (see Vlek and Stallen, 1980) and it summarizes the cognitive dimensions (see Slovic, Fischhoff and Lichtenstein, 1984; Vlek and Hendrickx, 1988).

Table 1: Formal definitions (upper part) and main cognitive dimensions (lower part) of risk

---

a. probability of a specified loss (or accident)

b. size (or seriousness) of a credible loss

c. expected loss: probability times size of loss

d. graph of probabilities and associated losses for a specific activity

e. weight of possible loss relative to weight of comparable gain

f. variance of probability distribution over all possible consequences

---

A. potential degree of harm or lethality

B. controllability through safety and/or rescue measures

C. number of people simultaneously exposed

D. familiarity of consequences and effects

E. degree of voluntariness of exposure

The crucial question here is: which definition and/or which (set of) cognitive dimension(s) is most effective in discussing and achieving 'safety'? With regard to both the formal definitions and the cognitive dimensions we can clearly see distinct references to the likelihood (credibility, possibility) of a loss or accident, and to its size or seriousness once it would occur. Formal definition b is reflected in cognitive dimensions A and C combined. More intriguing is the suggested correspondence between formal definition *a* and cognitive dimensions *B* and *D*: the *perceived* probability of a loss or accident seems to hinge upon the (perceived) controllability of a technological activity and, perhaps less so, upon one's familiarity with its possible consequences and effects.

Focussing on the probability, credibility or likelihood side of risk (and taking its effect side for granted), we follow Howell and Burnett (1978), Kahneman and Tversky (1982), and Vlek and Hendrickx (1988) in proposing that much of the controversy concerning the meaning and measurability of probability can be understood by pointing at the existence of basically different kinds of uncertainty, depending upon two underlying information dimensions. The first dimension is the relative availability of statistical or frequentistic data on past losses or accidents; this dimension implies that risk judgments are based on relative frequency information and/or on (cognitive) scenario information, depending upon which source is most available or most easily constructed.

The second dimension is defined in terms of the degree to which the likelihood of an unwanted event depends upon external (environmental) factors ('out there') or is dependent upon internal (human) factors associated with skills, knowledge and handling experience. Table 2 provides a sketch of the taxonomy of uncertainties thus obtained.

Table 2: Four basically different kinds of uncertainty, depending upon two information dimensions; the cells describe the data on which probability or risk judgments would be based.

	FREQUENTISTIC	NONFREQUENTISTIC
EXTERNALLY CONTROLLED	a. observed relative frequencies of 'fortuitous' events	b. scenarios, plans, properties or propensities of systems, objects or materials
INTERNALLY CONTROLLED	c. observed (calibrated) relative frequencies of success and failure	d. belief in feasibility of new activity or operation, or in truth of hypothesis

The horizontal dimension of Table 2 obviously refers to different uses and interpretations of the probability concept. Here, the message is that there may be situations in which there just is no frequentistic data-base and one has to go by scenario information. This very often happens in everyday personal risk taking and it raises questions about how people construct, store and retrieve mental scenarios pertaining to risky activities which they have experienced more or less often.

The vertical dimension of Table 2 refers to the well-known distinction between uncontrollable technical and environmental risk factors on the one hand, and controllable, but 'unpredictable' human factors on the other. Here we suggest that classical decision theory - implying a 'gambling view' of risk management - may be more applicable to the upper row concerning 'externally controlled' events. For dealing with activities having 'internally controlled' possible consequences, as indicated in the lower row, we would prefer to rely on a 'control view' of risk management (see further Vlek and Hendrickx, 1988).

The perspective offered in Table 2 has significant implications for risk assessment and management. *First*, what risk is and how it is best dealt with, depends upon the available kind of probabilistic information.

It, of course, also depends upon the nature of the relevant consequences - losses or accidents. Up to now, the *a priori* evaluation of possible accidents has been conducted by regulatory agencies in a rather simplistic manner (e.g., "the negative value of n possible deaths equals n-squared"). What is greatly missing is a multi-attribute modelling approach towards the evaluation of the varied loss, harm and damage caused by a major accident, prior to its (improbable) occurrence.

Together, these considerations imply that the risks of rather different activities, such as mountain climbing, car driving and electricity production, can not be compared other than on abstract statistical grounds. Such comparisons are rather meaningless (but they can be amusing) to people actually concerned about possible negative consequences of a specific activity.

Second, adequate risk management is a matter of organizing and maintaining a sufficient degree of (dynamic) control over a technological activity, rather than of continually, or just once, measuring accident probabilities and distributing the message that these are, and will be, 'negligeably low'. Thus, more often than not, 'acceptable risk' means 'sufficient control' and the latter has to be believed in by both the controlling agency and everyone else who is potentially affected by any negative consequences.

Third, the 'adequacy' or sufficiency of risk management as described above is a matter of regular review and judgment, preferably by a balanced and independent agency or institution. Such review and judgment might be based on specified technical requirements, the observance of valid regulations, the quality of organization and the amount of training and experience of responsible personnel. It is very hard indeed to set quantitative standards for testing the controllability of a technological project, just as it is hard to observe standards of sufficient control in everyday personal activities. In principle, however, this does not seem impossible, provided that one has a valid model of the relevant technological system, which specifies an operational set of controllability variables.

What seems to be less useful is to estimate the risk of a major accident and test this against some 'acceptable risk' standard. Such a strategy is rather limited in *practice*, because the various formal risk definitions in Table 1 are highly demanding, either with respect to the informational basis for probability assessment, or with regard to the relative weighting of possible losses as against gains.

The strategy is also limited *theoretically*, since from a decision-theoretic point of view testing against standards amounts to following a satisficing decision rule: consider a given choice alternative by itself and accept it when its scoring profile as regards an a priori set of criterion variables exceeds your profile of acceptance levels. On the one hand, such a satisficing model prevents one from performing an optimization analysis in an attempt to arrive at the best possible alternative. On the other hand: why should a satisficing decision rule be restricted to just considering *risk* acceptance standards? For constructive social decision making on technological projects it is essential to allocate "due weight to social and political factors when developing a 'formula' for assessing risks" (Pollak, 1985, p. 92).

A final point, which may help explain public scepticism about risk acceptance standards, concerns the accumulation of technological risks. When, in a particular LST siting case, regulators claim its acceptability on the basis of 'negligeably low risk' estimates, it is often overlooked that this specific LST projects comes in addition to other, existing technologies. With respect to nuclear power, for example, a common figure is the 'safe' requirement of a serious reactor accident to be less likely than once in 10,000 reactor years. However, given the 500-odd nuclear power stations currently in operation all over the world, it is easily calculated that the probability of a serious nuclear accident *anywhere*, and within the next *ten* years, approximates 40 percent. Nonexpert publics, generally exposed to a variety of low-probability technological risks, are sensitive to this cumulative exposure, and this contributes to their broader attitudes towards technological development (cf. Van der Pligt, this volume).

3. THE ROLE OF EXPERTS

Perhaps the most significant issue concerning expert opinion on technological projects is that there are different experts who - on valid scientific and/or technological grounds - may disagree with each other. A second observation is that socio-technical decision problems often are complex enough to make experts feel uncertain and have them come up with 'band-width' statements like "the estimated accident probability is in the order of 10 to the minus 4".

A natural consequence of expert disagreement is that nonexperts are going to doubt the existence of sufficient valid knowledge concerning the project under consideration. And these nonexperts may ultimately be correct when concluding that "science evidently has not yet formulated an answer to all the questions this technology is posing". The danger of expert ambiguity is that explicit statements may suffer from contamination by implicit policy preferences held personally by the expert or by his or her employing institution. When in doubt it is difficult not to sway gently in the direction of one's preferred conclusion. Especially in complex decision situations it is hard to keep facts and values properly separated.

The variety of different experts concerning a technological project and the ambiguity often brought about by the project's complexity make it a crucial question *which* experts are invited or allowed to conduct supportive research, to offer advice and/or to evaluate certain parts of the project. In Brouwer's contribution concerning the long-term surface storage facility for radioactive wastes, it appears that the Dutch government is relying on an independent experts' commission for the evaluation of an environmental impact statement (EIS) produced by the project's initiator.

The experiences with the operation of the Dutch EIS commission so far indicate that its judgment may be politically influential despite (or perhaps rather: thanks to) the fact that the government itself retains the responsibility for the definite evaluation of the EIS and for an eventual licensing decision. These experiences also suggest that for an independent experts' commission to be taken seriously, it is essential to stick to the expert's business: discussing factual matters in a politically detached way, and passing a balanced judgment on the environmental impact statement (and just that) prepared by a party which definitely has a stake in the proposed project.

A final question about experts is: what constitutes an expert? When, under what circumstances, and concerning which issues are the propositions and judgments of someone considered to be an 'expert' more trustworthy than those of any other nonspecialized but reasonably educated person? Perhaps the answer to these questions often is all too easily given, based as it may be on some kind of social prejudice ("trust the expert") concerning the role learned men and women may play in guiding our lives. Research shows, however, that different experts' judgments concerning the same topic may vary widely, that experts may err substantially, and that these phenomena tend to occur especially in complex, ambiguity-laden decision situations (see Shanteau, 1988).

The above considerations may provide grounds for a less reverent view of experts. They also imply an advice of modesty for experts. With respect to the case of radioactive waste storage Brouwer (this volume) remarked during a discussion that "the self-assuredness and technical arrogance of experts is doing a lot of harm to public opinion". This might only be subscribed to by the relevant experts themselves when they have accepted the view that social decision making about technological projects demands something like a multi-stage, multi-party procedure encompassing information exchange, analytical judgment and balanced deliberation.

4. QUALITY OF INDIVIDUAL JUDGMENT AND DECISION MAKING

In the first chapter we noted that social decision making about technological projects involves both individual judgment and social interaction and aggregation. Individual participants in the social process have to form their own representation of the problem, they have to appraise benefits, costs and risks from their own perspective (as well), and they have to accept and adhere to a rule that yields a decision in coherence with their various considerations. Under what circumstances can, and will, people do this optimally? And what conditions are likely to reduce individual capacities and motivation for rational information processing and decision making?

Behavioral decision research over the past few decades has increasingly revealed the serious limitations and shortcomings of classical decision theory for understanding and improving upon realistic decision processes (Kahneman, Slovic and Tversky, 1982; Pitz and Sachs, 1984; Hogarth, 1985; Slovic, Lichtenstein and Fischhoff, 1988). If a decision maker knows what (s)he is talking about (representational rationality), if that decision maker knows full well what (s)he wants and is able to express this consistently (goal-value rationality), and if (s)he follows logical assessment methods and a decision rule (procedural rationality), then we may speak of rational decision making. However, in less clearly developed situations one always faces the question of how far the decision analysis should be taken and to what extent 'compositional' (instead of decompositional or analytical) rationality (Toda, 1980) would require one to stop thinking, deliver a (preference) judgment and *act* before the available decision time runs out.

Psychological decision research has yielded the widely accepted conclusion that the behavior of unaided 'laboratory' decision makers often significantly deviates from the prescriptions of normative models like expected utility maximization or Bayesian revision of probabilities. Such decision makers would be plagued by judgmental distortions and biases showing that contextual factors, the particular assessment method used and the way a decision problem is 'framed' may lead to overall preference reversals.

We do not know, however, under what particular conditions people are prone to such 'irrational' phenomena, and under what circumstances (e.g., a complex dynamic situation) the underlying cognitive processes would have a certain functional value, or which conditions would be favorable toward *un*distorted judgment and decision making. We do know that stress may be rather detrimental to the quality of human information processing and judgment, especially the stress coming from complex problems that have to be solved under a fair amount of social and time pressure (see, e.g., Janis and Mann, 1977). And we know that human willingness to take risks increases, the further in the future a negative outcome of a decision is likely to occur (Björkman, 1984).

In controversial debates about the acceptability of new technology one may often observe that the parties involved hold on to different problem definitions and to different basic values relevant for evaluating choice alternatives. This might be obvious on first sight, but the question should be asked how one could - initially - *have* a clear problem definition and a reliable preference order when the decision problem at hand is fairly complex and when the feasible options are rather ambiguous with respect to their possible future consequences and effects.

Would not such a type of decision problem almost 'naturally' evoke an inclination to sit back, observe, analyze, weigh and aggregate, in order to come up with a final statement of preferences that through the very way it has been arrived at could not possibly be very outspoken? Is a rapid and outspoken position concerning a particular technological project a sign of unthoughtful ('irrational, emotional') decision making; does it indicate that 'deep personal values' are being touched upon; or is it to be seen as a reaction to the supposedly 'unthoughtful' preferences other involved parties may express at a premature moment in time?

A well designed social decision procedure for technological projects could properly deal with both the problems of biased (stress-affected) information processing and of too (early) outspoken problem definitions and policy preferences. On the one hand it should provide the conditions for careful, multi-faceted and balanced information processing and judgment. On the other hand it should support participants in *constructing* valid and reliable final preferences where these did not exist (or literally had 'no reason' to exist) beforehand.

The fact that an explicit method or procedure is invoked to assist one in constructing initially ill-formed preferences, implies that a strong emphasis is to be put on the quality of that decision method or procedure. It also implies that explicit decision aids are most helpful when initial policy preferences are vague and shaky, and apparently in need of more solid foundations and a clearer structure.

Probably little help of decision aids is to be expected in a polarized social situation in which different parties have already taken 'uncompromising' stands. Under such circumstances a decision-theoretic framework could at most be used party-wise, to chart the differing views, expectations and value judgments (much like Chen and Mathes, this volume, have done this). Sensitivity analysis - the powerful and 'fun' weapon of the decision analyst - might then be employed to show to each party 'what it matters' for their final preference order. Other, less formal techniques (for instance, just plain exposure to one another's worlds of thought) might be needed to cool off a polarized social situation.

5. COLLECTIVE DECISION MAKING

There are three basic types of 'social' decision problem:
  (1) collective decision making aimed at the specification of a common final preference order,
  (2) interdependent (e.g., competitive or cooperative) individual decision making having collective consequences, and
  (3) individual (or institutional) decision making aimed at a certain social distribution of benefits, costs and risks.
The first and second types of problem have been separately discussed by Bezembinder, and Wilke (both this volume), respectively. The third type of problem is touched upon by several authors throughout this volume; in essence it is the problem of the benevolent dictator who wishes to be fair to his subjects. Otherwise the procedures and behavior falling under the first and second types of problem above are instrumental in achieving fairness in the distribution of welfare.

Figure 1:The trilemma of social choice represented by three principles (solid triangle) and three practical approaches (dashed triangle) each meeting only a pair of them.

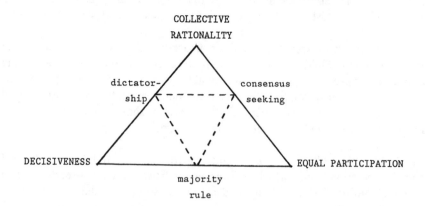

Bezembinder's treatment of collective choice theory (this volume) makes it clear that perfectly rational, egalitarian *and* 'decisive' rules for collective decision making can not be formulated. In fact from his and Blair and Pollak's (1983) exposition we may infer Figure 1 which represents the 'trilemma of social choice': three cherished principles (capital letters) accompanied by three practical approaches (lower-case letters), where each of the latter violates one (the opposite) principle.

The impossibility of formulating a collectively rational decision rule that fulfills reasonable democratic principles adds to the limited rational theory available to model *individual* decision making in ill-defined situations (see Section 4). Together they emphasize the importance of using a *fair procedure* for deciding about technological projects. And this inevitably must be a compromise on (collective) rationality, decisiveness and equal participation of those involved.

To avoid the use of power - which 'should be the last resort', in Bezembinder's (this volume) view - a fair social decision procedure should begin with finding an 'efficient set' of choice alternatives (where no option is dominated by any other), and it should promote a climate in which participants express social and/or ethical values rather than their own individual preferences. Dienel (this volume) reports that in the German 'planning cell' experiments this seems to have been achieved.

Interdependent individual decision behavior having collective consequences often yields a social dilemma situation: "if a larger number of individuals made 'cooperative' choices we would not have such an undesirable collective result". Wilke (this volume) points out that true social dilemmas are hard to resolve, except by the use of power (on the usefulness of which he disagrees with Bezembinder) or by changing the nature of the social decision situation such that it stops being a true dilemma (e.g., by introducing rewards for cooperative choices and/or punishments for defective ones). In a recent empirical paper Dawes, Van de Kragt and Orbell (1988) additionally stress the importance of group identity in bringing about cooperative choice behavior. Wilke proposes to combine 'soft' and 'hard' management approaches for solving important social dilemma problems; see his Figure 1.

There seems to be wide agreement among various authors that social conflicts and dilemmas may best be resolved using some kind of analytical approach. Vári (this volume) argues that conflict resolution either by accommodation, bargaining or creative confrontation, may profit from a decision-theoretic framework for identifying the different kinds of conflict. An analytical decision procedure enables one to lay out the components of a decision problem. It may illuminate the cognitive sources of social conflict, in terms of differing information variables and weights being attributed to them (Brehmer, this volume). It may help one to formulate a commonly acceptable 'requisite problem representation' (Phillips, this volume). It allows one to see which expectations or value judgments are crucial in determining a final preference order. It provides a framework for charting and elaborating individual parties' views and evaluations of a decision problem (Chen and Mathes, this volume).

Through these various characteristics, an analytical social decision approach constitutes a way to get around direct (personal) preference confrontations. Instead, it offers a procedure and methods to gradually develop and construct reliable social or ethical preferences concerning complex choice alternatives, whereby the initial personal preferences may change in the process.

Of course, analytical social decision methods would hardly get a chance in a technological emergency situation. As Rosenthal and Van Duin (this volume) point out, emergency conditions provoke centralized bureaucratic decisions based on sympathetic information sources, and often taken by analogy to previous similar situations. This is rather different from the kind of social decision making favored here; perhaps striving for balanced analytical decisions in an emergency situation is too much asked for. Nevertheless one wonders to what extent 'emergency preparedness' could also involve the availability of efficient decision methods that would exclude or diminish some of the nastier errors authorities can make under stressful socio-technical conditions.

6. COMPUTER PROGRAMS FOR SOCIAL DECISION SUPPORT

The chapter by De Hoog, Breuker and Van Dijk (this volume) reveals that a lot of attention has recently been devoted to the development of computer programs for decision support. In comparison with laboratory tests of decision theory, the design of practical decision aids is the real litmus test for the validity of prescriptive decision models. Workers in this area continually face the prospective client's question of whether the proposed program actually helps, and to what extent its utilization yields 'better decisions'. There are several important issues to be faced before one adopts computer-programmed decision aids in connection with technological projects.

To the extent that analytical decision making comprises problem structuring as well as evaluation of choice alternatives, an ideal programmed decision aid should incorporate: (a) an adequate data- or knowledge-base, (b) a set of models appropriate for structuring various types of decision problem, and (c) coherent procedures for arriving at optimal decisions. In decision aiding, content, structure and procedure all are essential ingredients (Phillips, this volume).

The construction of a valid and efficient data- and knowledge-base for a particular class of decision problems is rather difficult and expensive due to the required specifications. Therefore such 'expert-systems' are still rare and can only survive in working environments (e.g., medical hospitals) where a large number of similar decision problems both provide the necessary underlying statistics and make generalized applications possible at all.

For these reasons procedural decision aids are currently more popular and more numerously available. This category of aids is based on formal models from rational (individual) decision theory, such as the expected utility and multi-attribute utility models. These - as De Hoog et al.'s program PANIC demonstrates - may be adapted for use in group settings; POLICY, discussed by Brehmer in this volume, is another example. Procedural decision aids capitalize on coherent assessment and aggregation of the decision maker's expectations and preferences, which they help elicit in accordance with a general structural model (usually an options by attributes or an options by events matrix).

Formal models to represent decision problems so far constitute only a limited set. Taxonomic work is needed that could lead to a contingency model of decision-aiding, with various methods, techniques and decision rules being optimally allocated to basically different kinds of decision problems.

Operational criteria for establishing the 'goodness' of a decision advised through a computer-programmed method are not easily given. The validation research necessary to specify performance profiles which could convince prospective users is still scarce; such research is difficult to design anyway. The most convincing criterion for decision goodness, of course, is the attractiveness of the decision's actual *outcome(s)*. However, this would be a post-hoc criterion and a tricky one at that, since the outcome may not occur soon enough, it may be elusive, or it may be co-determined by uncontrollable external factors (the decision maker's luck or lack of it).

There seem to be three kinds of feasible criteria for evaluating decision goodness *a priori*: (1) rationality or coherence of process, (2) intuitive-emotional acceptance of, or trust in, the 'selected' alternative, and (3) user-satisfaction with the method followed. Theoretically, only the first kind of criterion is sound enough. However, given the limitations of available decision theory for modelling practical (social) decision procedures one is bound to look also at more descriptive criteria.

In fact, accepting the 'best' alternative resulting from a computer-programmed decision method seems to be an iterative matter of looking into the analytical method and consulting one's intuitive feelings, meanwhile spiralling up to a convincing convergence of the two. Analysis is followed up by synthesis which then is corroborated by further analysis, et cetera, until your intuition becomes 'one' with the (result of the) analysis. Perhaps this is the way in which computer-supported decision making, either by individuals or in groups, could best be conducted.

Computer-programmed decision aids seem potentially very useful to support the kind of analytical social decision procedure discussed by various contributors to this volume. The charting of individual parties' views, the pinpointing of elemental sources of conflict, the stepwise aggregation of opinions and preferences, and, above all, the rapid execution of sensitivity analyses, this can all be done much faster and more smoothly with the help of a well-programmed computer.

This presupposes that theoretically sound and operationally flawless computer software is available which eventually yields a high score profile on the evaluation criteria discussed above. Such software, and flexible hardware to run it on, could certainly make field experiments on citizen participation, like the ones conducted by Chen and Mathes, and by Dienel (both this volume), more feasible and less burdensome from an administrative point of view. A stimulating perspective and various useful ideas about group decision support using computers, are offered by DeSanctis and Gallupe (1987); see also Gray (1986).

7. THE NEED FOR INCREASED PUBLIC INVOLVEMENT

With the increase of scale in the realization of technological projects there naturally arose a growing demand for public information and independent (government) inspection and regulation. Responsible authorities in many cases have gone as far as arranging for public hearings to be held. In other cases they have been dodging this problem, apparently supposing that otherwise the planned decision making procedure would be irreparably damaged. And if authorities, or the initiators of a project, do try to communicate with members of the public, "experience repeatedly demonstrates that a lack of early and continuing involvement is a characteristic source of failure for public participation programs" (Kasperson, 1986, p. 277).

Here are some reasons for the inevitability of multi-party, representative participation in decision making about technological projects:

1. The multidimensionality of the actual and possible consequences of a technological project implies that a variety of different groups and organizations is involved. A valuable social principle stipulates that those (potentially) affected should also be involved in the relevant decision making procedure.
2. The complex, ill-defined and 'social' character of a large-scale technological project defies an 'objective, rational decision' guided by unambiguous expert knowledge. Thus a balanced, multi-party and multi-expert analysis and evaluation are required, in order to achieve a necessary degree of social accommodation.
3. The exposure of various public groups to involuntary and personally uncontrollable technological risks requires for its acceptance an extraordinary faith in the responsible initiators, operators and regulators; see also Starr (1985). This confidence is gnawed at, every time an expert 'assures' nonexperts that things are safe enough without letting those nonexperts take a look in his or her kitchen.
4. Through the complexity and imperfect controllability of its possible consequences, large-scale technology itself continually weakens public faith in experts ("how could *they* know all this thoroughly enough?"). Expert statements in themselves are insufficient to achieve a public belief that the technology is adequately

controllable. (Public trust is further damaged by suggested or reported incidents, without any serious accident ever occurring.)

The call for increased public participation in decision making concerning technological projects has been, and is being variously answered. From the policy maker's point of view, a list of escalating possibilities is as follows.
* "we'll tell them about it";
* "we'll listen to what they have to say about it";
* "we'll compensate them for it";
* "we'll give them a chance to judge for themselves";
* "let's get together and evaluate this collectively";
* "let's form a decision-and-control board that represents us all".

Several of these items could fall under so-called *risk communication* programs. The provision of public information about technological projects is a formidable problem in itself, to be attacked from various possible theoretical models (see Cvetkovich, Vlek and Earle, this volume). We believe, however, that optimizing information provision about technological risks is only part of the solution to problems of social controversy. At the heart of such problems, it seems, lies an unfulfilled public desire to know more, to have greater decision power, and above all, to have increased (either perceived or actual) control over the consequences of technological developments.

Supporting this point of view is the interesting correspondence of Hisschemöller and Midden's (this volume) Technocratic, Market, and Distributive-Justice policy approaches, respectively, with Cvetkovich et al.'s (this volume) Factual Information, Personal Gain/Loss and Values models for designing hazard communication programs. A further suggestion is that Hisschemöller and Midden's Public Participation approach, if appropriately conducted, should be compatible with Cvetkovich et al.'s communication model for Risk-Adaptive Decision Making. Not surprisingly, different decision approaches towards technology siting may very well go along with different models for informing the public about technological hazards.

8. TOWARDS A PROJECT-ORIENTED SOCIAL DECISION PROCEDURE

In typical cases the actual consequences of a socio-technical decision are hard to predict, difficult to control, and they may variously affect different social groups and organizations. Under such conditions a high-quality decision *procedure* is about the only thing which may cause the decision makers to believe that these consequences might come close to what they currently desire to achieve. This message has been relayed in several previous chapters and it also comes out of more specific discussions about the notion of 'a good decision' (Edwards, Kiss, Majone and Toda, 1984).

With respect to socially acceptable decision making about LST projects several principles have recently been advanced, e.g., by Pollak (1985) and Kasperson (1986). Also, a number of general approaches have been distinguished (see Hisschemöller and Midden, this volume). What seems to be lacking, however, is a more precise outline of a procedure that allows for diverse participation, is project-oriented, and is sufficiently well organized to deal with the individual judgment problem and the social interaction and aggregation problem as distinguished in the 'Introduction and Overview' chapter. Let us consider the general approaches first, then list a number of principles and conditions, before we sketch the outline of a project-oriented social decision procedure.

Hisschemöller and Midden's (this volume) four different approaches all seem to imply an important element for an effective social decision procedure: (1) expert technical analysis and judgment should be available (Technocratic approach); (2) those especially strained by a project may just wish to be compensated for this (Market approach); (3) potentially affected groups may wish to have their say and be taken seriously (Public Participation approach); and (4) the need for a fair social distribution of the benefits, costs and risks of a project may have to be acknowledged, at least in principle or as an ideal (Distributive-Justice approach).

Prior to the start of the Dutch 'Societal Discussion on (nuclear) Energy Policy' (see the 'Introduction and Overview' chapter) an "Initiative Group Energy Discussion" in 1978 published a list of seven guidelines for organizing this nationwide debate (see Vlek, 1986, p. 144). Considering these with regard to a more constrained project-oriented social decision procedure, we would select and slightly reword six of them, as follows.

1. Diverse participation right from the beginning of a decision making process enhances the quality of that process for all parties involved.

2. Problem definition, procedure, tasks, time schedule and responsibilities should be discussed and agreed upon in an early stage.
3. Apart from representative groups of the public concerned, special interest groups and organizations should also take part in the process.
4. Direct contacts between the participating groups and (eventually) responsible authorities should be established and maintained.
5. An independent, multidisciplinary and authoritative coordinating committee should be appointed, which is to conduct and oversee the process from beginning to end.
6. The quality of distributed information, expert advice and reports coming out of the process should be warranted through regular checks by the coordinating committee.

More recently, Pollak (1985, p. 92) has listed five conditions under which "dissenting groups are more likely to express their concerns in a constructive manner and to communicate effectively with administrative agencies":

   a. the allocation of due weight to social and political factors when developing a 'formula' for assessing risks;
   b. appropriate involvement of all main affected interests;
   c. unbiased management of the regulatory process;
   d. a fair distribution of expertise among affected parties;
   e. the decisions to be taken are not prejudged.

Pollak (1985, p. 89) also stresses "the creation of a framework of negotiation to allow for a reasonable possibility of achieving an agreement". To us this phrase means that a methodological framework is needed which would allow participants to effectively tackle the individual judgment problem and the social interaction and aggregation problem as distinguished in the 'Introduction and Overview' chapter.

From Dienel (this volume), finally, we borrow the requirement that participants in a social decision procedure should be given the time, the expert information and the instruments to perform their task in a suitable setting, so that social values are relied upon rather than individual values (see also Bezembinder's final section, this volume).

The above list of principles, requirements and conditions for effective social decision making about technological projects has led us to design the following outline of a 'recommended procedure'. In this proposal we have attempted to combine methodological elements from the earlier chapters.

Step 1.   Form an independent, multidisciplinary and authoritative Coordinating Committee (CC) to start off, stimulate and guide the decision making process.

Step 2.   Let this CC formulate an initial problem description which is then presented to identified groups and organizations (potentially) involved in the project.

Step 3.   Request the involved groups and organizations to react to the initial problem description by a specific response or (counter) proposal, accompanied by necessary background information from their side.

Step 4.   Let the CC review, sift and supplement the responses, proposals and information obtained, in order to:
   a. prepare a revised, possibly broader and more balanced description of the decision problem;
   b. prepare background materials needed to understand and evaluate possible decision alternatives; and
   c. revise the list of parties sufficiently involved (potentially) to be playing an active role in the remainder of the process.

Step 5.   From the general public concerned and by random sampling, set up a number of (paid) 'citizen groups' of about fifteen persons each (cf. Dienel's planning cells, this volume), and ask them to study (individually) the revised problem description together with the background materials.

Step 6.   Set up a number of (paid) 'organizational groups' composed of delegates from special interest groups, organizations and/or agencies (potentially) involved in the decision problem. Each organizational group should be made up of about ten people. Ask them similarly to study the revised problem description and the prepared background materials.

Step 7.   Let each citizen group and each organizational group go through the first of two successive 'decision conferences' (cf. Phillips, this volume) separated by an intermission of several weeks or months. During this first, *problem structuring* conference, a decision-analytic representation of the problem is aimed for. This should include a description of the existing and, perhaps several, desired states, a list of feasible policy options including the original project, and a reasonably detailed description of the possible consequences and effects of the various options. For the latter, scenarios may have been, or have to be, constructed (see Jungermann, 1985, for a relevant overview). To the extent that sufficient agreement exists, a preliminary evaluation of the feasible options may be attempted. The output of each group's deliberations is reported back to the CC.

Step 8. During the 'intermission' the CC collects the output of every group's first decision conference, formulates an overall representation of the decision problem as seen by the various groups, and it thereby indicates 'unique' views on the problem or on specific components thereof. The CC also obtains the additional needed information and expert judgments requested by the different groups. Expert advice is obtained from different sides and summarized in a balanced way before it is distributed. The CC feeds the overall problem representation and the requested supplementary information back to each group.

Step 9. During the second, *evaluation and choice* conference, every group systematically evaluates each of the feasible options, including the ones believed relevant by only few groups, by its possible consequences and effects. This can best be done using an 'expected multi-attribute utility' framework. Herein, the value of a consequence is based on multiple aspects, and its contribution to the overall value of an option is weighted by its estimated probability of occurrence (see Phillips, Chen and Mathes, Lubbers, this volume). Alternatively - and suboptimally - available policy options could be assessed and tested against an adopted set of acceptability criteria, defined on a number of relevant variables. The result of this second decision conference - which builds on that of the first - is recorded in a report from each group; the report also contains a section with the group's comments on the procedure followed and the methods used.

Step 10. The CC collects and puts together the reports from all groups, and it prepares a review listing the various groups' preferences or preference orders. The CC thereby reveals the main agreements and (remaining) disagreements both within and between the groups, as reported. Depending upon the amount of agreement the CC may formulate one or more policy recommendations supported by the outcomes of all groups' decision conferences. The CC's summary report is fed back to all participating groups, and it is presented to the government authority responsible for a final decision.

The procedure just outlined obviously has to be detailed, and it may either be constrained or expanded, depending upon the nature of the technological project considered. For practical operationalizations one would have to answer particular questions, e.g., about the number of citizen and organizational groups to be formed, the role of group coordinators and reporters, specific information exchange among groups, varied group consultation of experts, the utilization of computer programs for decision support, and the accommodation of government authorities to the decision process as it unfolds.

More general questions pertain to the desirable level of education and motivation of participants, the duration and costs of the procedure, and the specific techniques for conflict identification and resolution (Brehmer, Vári, this volume) to be used during controversial episodes in the process.

The proposed procedure, therefore, may vary from case to case, after it has been detailed and adapted to the current circumstances of an envisaged technological project and to the prevailing administrative culture. It would, for example, certainly be accommodated differently to the various regulatory styles distinguished by Otway (1985, p. 4/5): the adversarial (U.S.A.), the consensual (U.K., Sweden), the authoritative (France) and the corporatist (West Germany) models, respectively.

In essence, however, the ten-step procedure is to satisfy most of the principles, requirements and conditions listed earlier in this chapter. In particular, it is aimed at balanced and competent coordination, diverse participation, expert- and information-supported decision analysis, and the final reporting of 'socially responsible' preferences about feasible policy options. The procedure thereby goes far beyond any of the more restricted approaches, especially a classical - 'undemocratic' - techno-cratic approach and a broad - but 'incompetent' - public participation approach.

The proposed procedure thus may support the political-administrative process dealing with large-scale technological projects, without taking the ultimate responsibility away from mandated government authorities. The costs of the procedure in terms of human effort, time and money may therefore be paid back very well by a deeper public understanding of large-scale technology and a greater public trust in the quality of the relevant decision making processes.

REFERENCES

Bezembinder, Th. (this volume). Social choice theory and practice.

Björkman, M. (1984). Decision making, risk taking and psychological time: review of empirical findings and psychological theory. *Scandinavian Journal of Psychology, 25*, 31-49.

Blair, D.H. and Pollak, R.A. (1983). Rational collective choice. *Scientific American, 249(2)*, 76-83.

Brehmer, B. (this volume). Cognitive dimensions of conflicts over new technology.

Brouwer, H.C.G.M. (this volume). Current radioactive waste management policy in the Netherlands.

Chen, K. and Mathes, J.C. (this volume). Value oriented social decision analysis: a communication tool for public decision making on technological projects.

Collingridge, D. (1980). *The social control of technology*. London: Frances Pinter.

Cvetkovich, G., Vlek, Ch. and Earle, T.C. (this volume). Designing technological hazard information programs: towards a model of risk-adaptive decision making.

Dawes, R.M., Van De Kragt, A.J.C. and Orbell, J.M. (1988). Not me or thee but we: The importance of group identity in eliciting cooperation in dilemma situations: Experimental manipulations. *Acta Psychologica, 68*, 83-98.

De Hoog, R., Breuker, E. and Van Dijk, T. (this volume). Computer assisted group decision making.

DeSanctis, G. and Gallupe, R.B. (1987). A foundation for the study of group decision support systems. *Management Science, 33(5)*, 589-609.

Dienel, P.C. (this volume). Contributing to social decision methodology: citizen reports on technological projects.

Edwards, W., Kiss, I., Majone, G. and Toda, M. (1984). What constitutes 'a good decision'? *Acta Psychologica, 56*, 5-27.

Gray, P. (1986). Group decision support systems. In: E. Mclean and H.G. Sol (Eds.), *Decision support systems: a decade in perspective* (pp. 157-171). Amsterdam: Elsevier North-Holland.

Hisschemöller, M. and Midden, C.J.H. (this volume). Technological risk, policy theories and public perception in connection with the siting of hazardous facilities.

Hogarth, R.M. (1985). *Judgment and choice; the psychology of decision.* 2nd Edition. Chichester/New York: Wiley.

Howell, W.C. and Burnett, S.A. (1978). Uncertainty measurement: a cognitive taxonomy. *Organizational Behavior and Human Performance, 22,* 45-68.

Janis, I.L. and Mann, L. (1977). *Decision making; a psychological analysis of conflict, choice and commitment.* New York: The Free Press.

Jungermann, H. (1985). Psychological aspects of scenarios. In: V.T. Covello, J.L. Mumpower, P.J.M. Stallen and V.R.R. Uppuluri (Eds.), *Environmental impact assessment, technology assessment and risk analysis; contributions from the psychological and decision sciences* (pp. 325-346). Berlin/Heidelberg: Springer Verlag.

Kahneman, D. and Tversky, A. (1982). Variants of uncertainty. *Cognition, 11,* 143-157.

Kahneman, D., Slovic, P. and Tversky, A. (Eds.).(1982). *Judgment under uncertainty, heuristics and biases.* Cambridge (U.K.): Cambridge University Press.

Kasperson, R.E. (1986). Six propositions on public participation and their relevance for risk communication. *Risk Analysis, 6(3),* 275-281.

Lubbers, F. (this volume). Planning and procedural aspects of the windfarm project of the Dutch Electricity Generating Board.

Otway, H. (1985). Regulation and risk analysis. In: H. Otway and M. Peltu (Eds.), *Regulating industrial risks; science, hazards and public protection* (pp. 1-19). London: Butterworths.

Perrow, Ch. E. (1984). *Normal accidents; living with high-risk technologies.* New York: Basic Books.

Phillips, L.D. (this volume). Requisite decision modelling for technological projects.

Pitz, G.F. and Sachs, N.J. (1984). Judgment and decision: theory and application. *Annual Review of Psychology, 35,* 139-164.

Pollak, M. (1985). Public participation. In: H. Otway and M. Peltu (Eds.), *Regulating industrial risks; science, hazards and public protection* (pp. 76-93). London: Butterworths.

Rosenthal, U. and Van Duin, M.J. (this volume). Decision making in technological emergencies.

Shanteau, J. (1988). Psychological characteristics and strategies of expert decision makers. *Acta Psychologica, 68,* 203-215.

Sjöberg, L. (1980). The risks of risk analysis. *Acta Psychologica, 45,* 301-321.

Slovic, P., Fischhoff, B. and Lichtenstein, S. (1984). Behavioral decision theory perspectives on risk and safety. *Acta Psychologica, 56,* 183-203.

Slovic, P., Lichtenstein, S. and Fischhoff, B. (1988). Decision making. In: R.C. Atkinson, R.J. Herrnstein, G. Lindzey and R.D. Luce (Eds.), *Stevens' handbook of experimental psychology. Volume 2* (pp. 673-738). New York: Wiley.

Starr, Ch. (1985). Risk management, assessment and acceptability. *Risk Analysis, 5(2),* 97-102.

Toda, M. (1980). Emotion and decision making. *Acta Psychologica, 45,* 133-155 (also in: M. Toda, *Man, robot and society.* The Hague: Martinus Nijhoff, 1981).

Van der Pligt, J. (this volume). Nuclear waste: public perception and siting policy.

Vári, A. (this volume). Approaches towards conflict resolution in decision processes.

Vlek, C.A.J. (1986). Rise, decline and aftermath of the Dutch 'Societal Discussion on (nuclear) Energy Policy' (1981-1983). In: H.A. Becker and A.L. Porter (Eds.), *Impact assessment today. Volume I* (pp. 141-188). Utrecht: Van Arkel.

Vlek, Ch. and Hendrickx, L. (1988). Statistical risk versus personal control as conceptual bases for evaluating (traffic) safety. In: J.A. Rothengatter and R.A. De Bruin (Eds.), *Road user behavior; theory and research* (pp. 139-151). Assen/Maastricht (Neth.) and Wolfeboro (N.H.): Van Gorcum.

Vlek, Ch. and Stallen, P.J. (1980). Rational and personal aspects of risk. *Acta Psychologica, 45,* 273-300.

Wilke, H.A.M. (this volume). Promoting personal decisions supporting the achievement of risky public goods.

AUTHOR INDEX*

---

* This Author Index does not contain (cross-)references to authors' contributions to the present volume. See especially the first and last chapters, and - of course - the table of contents.

Appendix: Authors' Addresses

Thom Bezembinder
University of Nijmegen, Institute for Cognition Research and Information Technology (NICI), Psychological Laboratory, P.O. Box 9104, 6500 HE Nijmegen, the Netherlands.

Berndt Brehmer
Department of Psychology, University of Uppsala, P.O. Box 1854, S-751 48, Uppsala, Sweden.

Eric Breuker
Social Science Informatics Group, University of Amsterdam, Herengracht 196, 1016 BS Amsterdam, the Netherlands.

Henk C.G.M. Brouwer
Ministry of Housing, Physical Planning and the Environment, Division of MER, P.O. Box 450, 2260 MB Leidschendam, the Netherlands.

Kan Chen
College of Engineering, 111 Technical Information and Analysis Laboratory, 2360 Bonisteel Boulevard, Ann Arbor, Michigan, 48109-2108, U.S.A.

George Cvetkovich
Western Institute for Social and Organizational Research, Department of Psychology, Western Washington University, Bellingham, Washington 98225, U.S.A.

Peter C. Dienel
Citizen Participation Research Unit, Wuppertal University, P.O. Box 100127, D-5600 Wuppertal 1, Federal Republic of Germany.

Timothy C. Earle
Western Institute for Social and Organizational Research, Department of Psychology, Western Washington University, Bellingham, Washington 98225, U.S.A.

Robert de Hoog
Social Science Informatics Group, University of Amsterdam, Herengracht 196, 1016 BS Amsterdam, the Netherlands.

Matthijs Hisschemöller
Department of Sociology, Erasmus University Rotterdam, P.O.Box 1738, 3000 DR Rotterdam, the Netherlands.

Frits Lubbers
Division of Physical Planning and Environmental Affairs, Dutch Electricity Generating Board (Sep), P.O.Box 9035, 6800 ET Arnhem, the Netherlands.

J.C. Mathes
College of Engineering, 111 Technical Information and Analysis Laboratory, 2360 Bonisteel Boulevard, Ann Arbor, Michigan, 48109-2108, U.S.A.

Cees J.H. Midden
Department of Psychology, Division of Energy and Environmental Research, University of Leiden, P.O. Box 9509, 2300 RA Leiden, the Netherlands.

Lawrence D. Phillips
Decision Analysis Unit, London School of Economics and Political Science, Houghton Street, London WC 2A 2AE, England.

Uriel Rosenthal
Department of Administrative Science, University of Leiden, Rapenburg 59, 2311 GJ Leiden, the Netherlands.

Joop van der Pligt
University of Amsterdam, Psychological Laboratory, Social Psychology Division, Weesperplein 8, 1018 XA Amsterdam, the Netherlands.

Tibert van Dijk
Social Science Informatics Group, University of Amsterdam, Herengracht 196, 1016 BS Amsterdam, the Netherlands.

Menno J. van Duin
Department of Administrative Science, University of Leiden, Rapenburg 59, 2311 GJ Leiden, the Netherlands.

Anna Vári
Hungarian Institute for Public Opinion Research, Akadémia u. 17, (P.O. Box 587), H-1054 Budapest, Hungary.

Charles Vlek
Department of Psychology, University of Groningen, Kraneweg 4, 9718 JP Groningen, the Netherlands.

Henk A.M. Wilke
University of Groningen, Department of Psychology, Kraneweg 4, 9718 JP Groningen, the Netherlands.

THEORY AND DECISION LIBRARY

SERIES A: PHILOSOPHY AND METHODOLOGY OF THE SOCIAL
            SCIENCES

Already published:

Conscience: An Interdisciplinary View
Edited by Gerhard Zecha and Paul Weingartner
ISBN 90–277–2452–0

Cognitive Strategies in Stochastic Thinking
by Roland W. Scholz
ISBN 90–277–2454–7

Comparing Voting Systems
by Hannu Nurmi
ISBN 90–277–2600–0

Evolutionary Theory in Social Science
Edited by Michael Schmid and Franz M. Wuketits
ISBN 90–277–2612–4

The Metaphysics of Liberty
by Frank Forman
ISBN 0–7923–0080–7

Towards a Strategic Management and Decision Technology
by John W. Sutherland
ISBN 0–7923–0245–1